Foundations of Matrices

Foundations of Matrices

Arun Dubey

Oxford Book Company
Jaipur ▪ New Delhi

Published By

Oxford Book Company

Regd. Off. "Bony Residency", Gate No. 2
Opp. Tilak Public School, Vishweswariya Nagar,
Gopalpura Road, Jaipur - 302018, India
Ph: 0141 - 2761280, Telefax: 2761381
email - oxfordbookjpr@gmail.com
website - oxfordbookcompany.com

Delhi Office: 102, 1st Floor, Satyam House,
4327/3, Ansari Road, Darya Ganj,
New Delhi - 110002, India
Ph: 011-45652440

Distributed By
KEYA INTERNATIONAL PRESS LLP
E-mail: keyainternationalpressllp@gmail.com

First published in India, 2026

ISBN : 978-93-5524-549-6

Typeset by Shivangi Computers, Jaipur

Printed and bound in India by Thomson Press Ltd.,
New Delhi

Contents

Preface

Matrices form the backbone of modern mathematics, seamlessly integrating into diverse disciplines such as physics, computer science, engineering, economics, and artificial intelligence. Their profound applications span from solving systems of linear equations to advanced topics like quantum mechanics, data science, and neural networks. Foundations of Matrices is designed to provide a comprehensive, systematic, and rigorous introduction to matrices, equipping students, researchers, and professionals with the necessary tools to master this fundamental concept.

This book offers a balanced treatment of theory and application, ensuring that readers not only understand the mathematical structures underlying matrices but also develop the analytical skills required to apply them effectively. Beginning with the basic properties of matrices, including types, operations, and determinants, the book gradually progresses to more advanced topics such as eigenvalues, eigenvectors, diagonalization, and singular value decomposition. Special emphasis is placed on computational techniques, numerical methods, and their relevance in real-world problem-solving.

A distinctive feature of Foundations of Matrices is its focus on conceptual clarity and logical development. Each topic is carefully structured, starting from fundamental principles and advancing towards more complex ideas in a stepwise manner. Detailed proofs, illustrative examples, and geometric interpretations are provided to reinforce understanding and foster intuitive learning. Furthermore, the book integrates modern computational tools and software applications, allowing readers to bridge theoretical knowledge with practical implementation.

Beyond its mathematical rigor, this book highlights the

significance of matrices in interdisciplinary research. From cryptography and image processing to optimization and machine learning, matrices play a crucial role in shaping contemporary scientific and technological advancements. By presenting both classical and contemporary perspectives, this book ensures that readers gain a holistic appreciation of matrices as both an abstract mathematical framework and a powerful tool for innovation.

Each chapter is supplemented with exercises ranging from fundamental practice problems to thought-provoking challenges, encouraging readers to develop a deeper grasp of the subject. The inclusion of research-oriented discussions and historical insights further enriches the learning experience, making this book a valuable resource for anyone eager to explore the vast potential of matrices.

Foundations of Matrices is intended to be a helpful companion on the path of mathematical exploration. Since mathematics is a constantly changing field that benefits from cooperation and ongoing growth, constructive criticism and ideas for enhancement are always appreciated.

— ***Arun Dubey***

Chapter 1

Introduction to Metrices

In mathematics, a matrix (plural matrices) is a rectangular array of numbers, symbols, or expressions, arranged in *rows* and *columns*. The individual items in a matrix are called its *elements* or *entries*. An example of a matrix with 2 rows and 3 columns is

$$\begin{bmatrix} 1 & 9 & -13 \\ 20 & 5 & -6 \end{bmatrix}.$$

Matrices of the same size can be added or subtracted element by element. The rule for matrix multiplication is more complicated, and two matrices can be multiplied only when the number of columns in the first equals the number of rows in the second. A major application of matrices is to represent linear transformations, that is, generalizations of linear functions such as $f(x) = 4x$. For example, the rotation of vectors in three dimensional space is a linear transformation.

If R is a rotation matrix and v is a column vector (a matrix with only one column) describing the position of a point in space, the product Rv is a column vector describing the position of that point after a rotation. The product of two matrices is a matrix that represents the composition of two linear transformations. Another application of matrices is in the solution of a system of linear equations. If the matrix is square, it is possible to deduce some of its properties by computing its determinant. For example, a square matrix has an inverse if and only if its determinant is not zero. Eigenvalues and eigenvectors provide insight into the geometry of linear transformations.

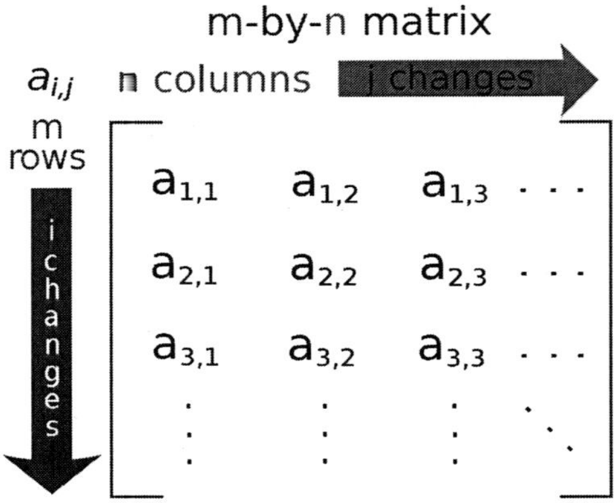

Figure: *Specific elements of a matrix are often denoted by a variable with two subscripts. For instance,* $a_{2,1}$ *represents the element at the second row and first column of a matrix A.*

Matrices find applications in most scientific fields. In physics, matrices are used to study electrical circuits, optics, and quantum mechanics.

In computer graphics, matrices are used to project a 3-dimensional image onto a 2-dimensional screen, and to create realistic-seeming motion. Matrix calculus generalizes classical analytical notions such as derivatives and exponentials to higher dimensions. A major branch of numerical analysis is devoted to the development of efficient algorithms for matrix computations, a subject that is centuries old and is today an expanding area of research. Matrix decomposition methods simplify computations, both theoretically and practically.

Algorithms that are tailored to the structure of particular matrix structures, e.g. sparse matrices and near-diagonal matrices, expedite computations in finite element method and other computations. Infinite matrices occur in planetary theory and in atomic theory. A simple example is the matrix representing the derivative operator, which acts on the Taylor series of a function.

Matrix (Mathematics)

In mathematics, a matrix (plural matrices) is a rectangular array of numbers, symbols, or expressions, arranged in *rows* and *columns*. The individual items in a matrix are called its *elements* or *entries*. An example of a matrix with 2 rows and 3 columns is

$$\begin{bmatrix} 1 & 9 & -13 \\ 20 & 5 & -6 \end{bmatrix}.$$

Matrices of the same size can be added or subtracted element by element. The rule for matrix multiplication is more complicated, and two matrices can be multiplied only when the number of columns in the first equals the number of rows in the second. A major application of matrices is to represent linear transformations, that is, generalizations of linear functions such as $f(x) = 4x$. For example, the rotation of vectors in three dimensional space is a linear transformation.

If R is a rotation matrix and v is a column vector (a matrix with only one column) describing the position of a point in space, the product Rv is a column vector describing the position of that point after a rotation. The product of two matrices is a matrix that represents the composition of two linear transformations. Another application of matrices is in the solution of a system of linear equations. If the matrix is square, it is possible to deduce some of its properties by computing its determinant. For example, a square matrix has an inverse if and only if its determinant is not zero. Eigenvalues and eigenvectors provide insight into the geometry of linear transformations.

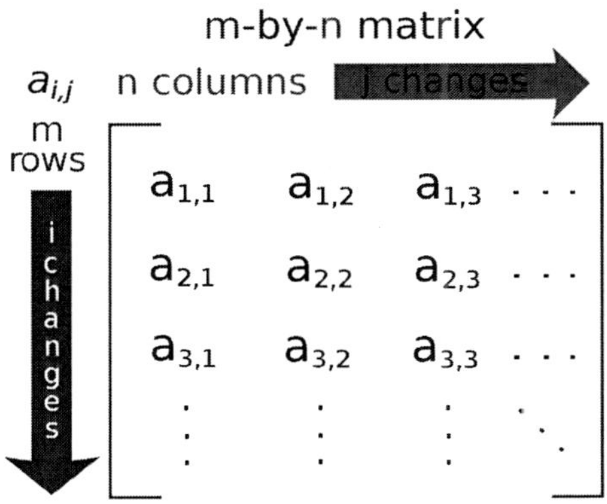

Figure: *Specific elements of a matrix are often denoted by a variable with two subscripts. For instance,* $a_{2,1}$ *represents the element at the second row and first column of a matrix A.*

Matrices find applications in most scientific fields. In physics, matrices are used to study electrical circuits, optics, and quantum mechanics. In computer graphics, matrices are used to project a 3-dimensional image onto a 2-dimensional screen, and to create realistic-seeming motion. Matrix calculus generalizes classical analytical notions such as derivatives and exponentials to higher dimensions.

A major branch of numerical analysis is devoted to the development of efficient algorithms for matrix computations, a subject that is centuries old and is today an expanding area of research. Matrix decomposition methods simplify computations, both theoretically and practically. Algorithms that are tailored to the structure of particular matrix structures, e.g. sparse matrices and near-diagonal matrices, expedite computations in finite element method and other computations. Infinite matrices occur in planetary theory and in atomic theory. A simple example is the matrix representing the derivative operator, which acts on the Taylor series of a function.

Definition

A *matrix* is a rectangular arrangement of mathematical expressions that can be simply numbers.

Commonly the m components of the matrix are written in a rectangular arrangement in the form of a column of m rows:

$$A = \begin{bmatrix} a_{11} & a_{12} & \cdots & a_{1n} \\ a_{21} & a_{22} & \cdots & a_{2n} \\ \vdots & \vdots & \ddots & \vdots \\ a_{m1} & a_{m2} & \cdots & a_{mn} \end{bmatrix}$$

For example,

$$\mathbf{A} = \begin{bmatrix} 9 & 13 & 5 \\ 1 & 11 & 7 \\ 3 & 9 & 2 \\ 6 & 0 & 7 \end{bmatrix}.$$

An alternative notation uses large parentheses instead of box brackets. The horizontal and vertical lines in a matrix are called *rows* and *columns*, respectively. The numbers in the matrix are called its *entries* or its *elements*. To specify the size of a matrix, a matrix with m rows and n columns is called an m-by-n matrix or $m \times n$ matrix, while m and n are called its *dimensions*. The above is a 4-by-3 matrix.

A matrix with one row (a $1 \times n$ matrix) is called a row vector, and a matrix with one column (an $m \times 1$ matrix) is called a column vector. Any row or column of a matrix determines a row or column vector, obtained by removing all other rows or columns respectively from the matrix. For example, the row vector for the third row of the above matrix A is

$$\begin{bmatrix} 3 & 9 & 2 \end{bmatrix}.$$

When a row or column of a matrix is interpreted as a value, this refers to the corresponding row or column vector. For instance one may say that two different rows of a matrix are equal, meaning they determine the same row vector.

In some cases the value of a row or column should be interpreted just as a sequence of values (an element of R^n if entries are real numbers) rather than as a matrix, for instance when saying that the rows of a matrix are equal to the corresponding columns of its transpose matrix. Most of this article focuses on *real* and *complex matrices*, i.e., matrices whose elements are real or complex numbers, respectively. More general types of entries are discussed below.

Notation

The specifics of matrices notation varies widely, with some prevailing trends. Matrices are usually denoted using upper-case letters, while the corresponding lower-case letters, with two subscript indices, represent the entries. In addition to using upper-case letters to symbolize matrices, many authors use a special typographical style, commonly boldface upright (non-italic), to further distinguish matrices from other mathematical objects. An alternative notation involves the use of a double-underline with the variable name, with or without boldface style..

The entry in the i-th row and the j-th column of a matrix is typically referred to as the i,j, (i,j), or $(i,j)^{th}$ entry of the matrix. For example, the (2,3) entry of the above matrix A is 7. The $(i, j)^{th}$ entry of a matrix A is most commonly written as $a_{i,j}$. Alternative notations for that entry are $A[i,j]$ or $A_{i,j}$. Sometimes a matrix is referred to by giving a formula for its

$(i,j)^{th}$ entry, often with double parenthesis around the formula for the entry, for example, if the $(i,j)^{th}$ entry of A were given by a_{ij}, A would be denoted $((a_{ij}))$.

An asterisk is commonly used to refer to whole rows or columns in a matrix. For example, $a_{i,-}$ refers to the ith row of A, and $a_{-,j}$ refers to the jth column of A. The set of all m-by-n matrices is denoted $\mathbb{M}(m, n)$.

A common shorthand is

$$A = [a_{i,j}]_{i = 1,...,m;\ j = 1,...,n} \text{ or more briefly } A = [a_{i,j}]_{m\times n}$$

to define an $m \times n$ matrix A. Usually the entries $a_{i,j}$ are defined separately for all integers $1 \le i \le m$ and $1 \le j \le n$. They can however sometimes be given by one formula; for example the 3-by-4 matrix

$$\mathbf{A} = \begin{bmatrix} 0 & -1 & -2 & -3 \\ 1 & 0 & -1 & -2 \\ 2 & 1 & 0 & -1 \end{bmatrix}$$

can alternatively be specified by $A = [i - j]_{i = 1,2,3;\ j = 1,...,4}$, or simply $A = ((i\text{-}j))$, where the size of the matrix is understood.

Some programming languages start the numbering of rows and columns at zero, in which case the entries of an m-by-n matrix are indexed by $0 \le i \le m - 1$ and $0 \le j \le n - 1$. This article follows the more common convention in mathematical writing where enumeration starts from 1.

Basic Operations

Matrix Addition

In mathematics, matrix addition is the operation of adding two matrices by adding the corresponding entries together. However, there are other operations which could also be considered as a kind of addition for matrices, the direct sum and the Kronecker sum.

Entrywise Sum

The usual matrix addition is defined for two matrices of the same dimensions. The sum of two $m \times n$ (pronounced "m by n") matrices A and B, denoted by A + B, is again an $m \times n$ matrix computed by adding corresponding elements:

$$\mathbf{A}+\mathbf{B} = \begin{bmatrix} a_{11} & a_{12} & \cdots & a_{1n} \\ a_{21} & a_{22} & \cdots & a_{2n} \\ \vdots & \vdots & \ddots & \vdots \\ a_{m1} & a_{m2} & \cdots & a_{mn} \end{bmatrix} + \begin{bmatrix} b_{11} & b_{12} & \cdots & b_{1n} \\ b_{21} & b_{22} & \cdots & b_{2n} \\ \vdots & \vdots & \ddots & \vdots \\ b_{m1} & b_{m2} & \cdots & b_{mn} \end{bmatrix}$$

$$= \begin{bmatrix} a_{11}+b_{11} & a_{12}+b_{12} & \cdots & a_{1n}+b_{1n} \\ a_{21}+b_{21} & a_{22}+b_{22} & \cdots & a_{2n}+b_{2n} \\ \vdots & \vdots & \ddots & \vdots \\ a_{m1}+b_{m1} & a_{m2}+b_{m2} & \cdots & a_{mn}+b_{mn} \end{bmatrix}$$

For example:

$$\begin{bmatrix} 1 & 3 \\ 1 & 0 \\ 1 & 2 \end{bmatrix} + \begin{bmatrix} 0 & 0 \\ 7 & 5 \\ 2 & 1 \end{bmatrix} = \begin{bmatrix} 1+0 & 3+0 \\ 1+7 & 0+5 \\ 1+2 & 2+1 \end{bmatrix} = \begin{bmatrix} 1 & 3 \\ 8 & 5 \\ 3 & 3 \end{bmatrix}$$

We can also subtract one matrix from another, as long as they have the same dimensions. A – B is computed by subtracting corresponding elements of A and B, and has the same dimensions as A and B. For example:

$$\begin{bmatrix} 1 & 3 \\ 1 & 0 \\ 1 & 2 \end{bmatrix} - \begin{bmatrix} 0 & 0 \\ 7 & 5 \\ 2 & 1 \end{bmatrix} = \begin{bmatrix} 1-0 & 3-0 \\ 1-7 & 0-5 \\ 1-2 & 2-1 \end{bmatrix} = \begin{bmatrix} 1 & 3 \\ -5 & -5 \\ -1 & 1 \end{bmatrix}$$

Direct Sum

Another operation, which is used less often, is the direct sum (denoted by $\oplus$). Note the Kroneker sum is also denoted $\oplus$; the context should make the usage clear.

The direct sum of any pair of matrices A of size $m \times n$ and B of size $p \times q$ is a matrix of size $(m + p) \times (n + q)$ defined as

$$\mathbf{A} \oplus \mathbf{B} = \begin{bmatrix} \mathbf{A} & \mathbf{0} \\ \mathbf{0} & \mathbf{B} \end{bmatrix} = \begin{bmatrix} a_{11} & \cdots & a_{1n} & 0 & \cdots & 0 \\ \vdots & \ddots & \vdots & \vdots & \ddots & \vdots \\ a_{m1} & \cdots & a_{mn} & 0 & \cdots & 0 \\ 0 & \cdots & 0 & b_{11} & \cdots & b_{1q} \\ \vdots & \ddots & \vdots & \vdots & \ddots & \vdots \\ 0 & \cdots & 0 & b_{p1} & \cdots & b_{pq} \end{bmatrix}$$

For instance,

$$\begin{bmatrix} 1 & 3 & 2 \\ 2 & 3 & 1 \end{bmatrix} \oplus \begin{bmatrix} 1 & 6 \\ 0 & 1 \end{bmatrix} = \begin{bmatrix} 1 & 3 & 2 & 0 & 0 \\ 2 & 3 & 1 & 0 & 0 \\ 0 & 0 & 0 & 1 & 6 \\ 0 & 0 & 0 & 0 & 1 \end{bmatrix}$$

The direct sum of matrices is a special type of block matrix, in particular the direct sum of square matrices is a block diagonal matrix. The adjacency matrix of the union of disjoint graphs or multigraphs is the direct sum of their adjacency matrices. Any element in the direct sum of two vector spaces of matrices can be represented as a direct sum of two matrices. In general, the direct sum of n matrices is:

$$\bigoplus_{i=1}^{n} \mathbf{A}_i = \mathrm{diag}(\mathbf{A}_1, \mathbf{A}_2, \mathbf{A}_3 \cdots \mathbf{A}_n) = \begin{bmatrix} \mathbf{A}_1 & \mathbf{0} & \cdots & \mathbf{0} \\ \mathbf{0} & \mathbf{A}_2 & \cdots & \mathbf{0} \\ \vdots & \vdots & \ddots & \vdots \\ \mathbf{0} & \mathbf{0} & \cdots & \mathbf{A}_n \end{bmatrix}$$

where the zeros are actually blocks of zeros, i.e. zero matricies.

NB: Sometimes in this context, boldtype for matrices is dropped, matricies are written in italic.

Kronecker Sum

The Kronecker sum is different from the direct sum but is also denoted by $\oplus$. It is defined using the Kronecker product $\oplus$ and normal matrix addition. If A is n-by-n, B is m-by-m and I_k denotes the k-by-k identity matrix then the Kronecker sum is defined by:

$$\mathbf{A} \oplus \mathbf{B} = \mathbf{A} \otimes \mathbf{I}_m + \mathbf{I}_n \otimes \mathbf{B}.$$

Matrix Multiplication

In mathematics, matrix multiplication is a binary operation that takes a pair of matrices, and produces another matrix. This term may refer to a number of different ways to multiply matrices, but most commonly refers to the matrix product. This article will use the following notational conventions. Matrices are represented by capital letters in bold, vectors in lowercase bold, and entries of vectors and matrices are italic (since they are scalars).

Scalar Multiplication

In mathematics, scalar multiplication is one of the basic operations defining a vector space in linear algebra (or more generally, a module in abstract algebra). In an intuitive geometrical context, scalar multi-plication of a real Euclidean vector by a positive real number multiplies the magnitude of the vector without changing its direction. The term "scalar" itself derives from this usage: a scalar is that which scales vectors. Scalar multiplication is different from the scalar product, which is an inner product between two vectors.

Definition

In general, if K is a field and V is a vector space over K, then scalar multiplication is a function from $K \times V$ to V. The result of applying this function to c in K and v in V is denoted cv.

Scalar multiplication obeys the following rules *(vector in boldface)*:

- Left distributivity: $(c + d)v = cv + dv$;
- Right distributivity: $c(v + w) = cv + cw$;
- Associativity: $(cd)v = c(dv)$;
- Multiplying by 1 does not change a vector: $1v = v$;
- Multiplying by 0 gives the null vector: $0v = 0$;
- Multiplying by -1 gives the additive inverse: $(-1)v = -v$.

Here + is addition either in the field or in the vector space, as appropriate; and 0 is the additive identity in either. Juxtaposition indicates either scalar multiplication or the multiplication operation in the field.

Scalar multiplication may be viewed as an external binary operation or as an action of the field on the vector space. A geometric interpretation to scalar multiplication is a stretching or shrinking of a vector. As a special case, V may be taken to be K itself and scalar multi-plication may then be taken to be simply the multiplication in the field. When V is K^n, then scalar multiplication is defined component-wise.

The same idea goes through with no change if K is a commutative ring and V is a module over K. K can even be a rig, but then there is no additive inverse. If K is not commutative, then the only change is that the order of the multiplication may be reversed, resulting in the distinct operations *left scalar multiplication cv* and *right scalar multiplication vc.*

General Definition

Left multiplication: The scalar multiplication of a matrix A and a scalar λ gives another matrix λ A of the same size as A. The entries of λ A are given by

$$\lambda(A)_{ij} = (\lambda A)_{ij} = \lambda A_{ij}$$

explicitly:

$$\lambda\mathbf{A} = \lambda\begin{pmatrix} A_{11} & A_{12} & \cdots & A_{1m} \\ A_{21} & A_{22} & \cdots & A_{2m} \\ \vdots & \vdots & \ddots & \vdots \\ A_{n1} & A_{n2} & \cdots & A_{nm} \end{pmatrix} = \begin{pmatrix} \lambda A_{11} & \lambda A_{12} & \cdots & \lambda A_{1m} \\ \lambda A_{21} & \lambda A_{22} & \cdots & \lambda A_{2m} \\ \vdots & \vdots & \ddots & \vdots \\ \lambda A_{n1} & \lambda A_{n2} & \cdots & \lambda A_{nm} \end{pmatrix}.$$

If we are concerned with matrices over a more general ring, then the above multiplication is the *left multiplication* of the matrix A with scalar λ.

Right multiplication

Similarly, the *right multiplication* is defined to be

$$(A\lambda)_{ij} = A_{ij}\lambda$$

When the underlying ring is commutative, for example, the real or complex number field, the two multiplications are the same. However, if the ring is *not* commutative, such as the quaternions, they may not be equal.

Examples

For a real scalar and matrix:

$$\lambda = 2, \quad \mathbf{A} = \begin{pmatrix} a & b \\ c & d \end{pmatrix},$$

$$2\mathbf{A} = 2\begin{pmatrix} a & b \\ c & d \end{pmatrix} = \begin{pmatrix} 2a & 2b \\ 2c & 2d \end{pmatrix} = \begin{pmatrix} a & b \\ c & d \end{pmatrix}2 = \mathbf{A}2.$$

For quaternion scalars and matrices:

$$i\begin{pmatrix} i & 0 \\ 0 & j \end{pmatrix} = \begin{pmatrix} -1 & 0 \\ 0 & k \end{pmatrix} \neq \begin{pmatrix} -1 & 0 \\ 0 & -k \end{pmatrix} = \begin{pmatrix} i & 0 \\ 0 & j \end{pmatrix}i.$$

Matrix Product (Two Matrices)

Assume two matrices are to be multiplied (the generalization to any number is discussed below).

If A is an $n \times m$ matrix and B is an $m \times p$ matrix, the result AB of their multiplication is an $n \times p$ matrix defined only if the number of columns m in A is equal to the number of rows m in B.

General Definition

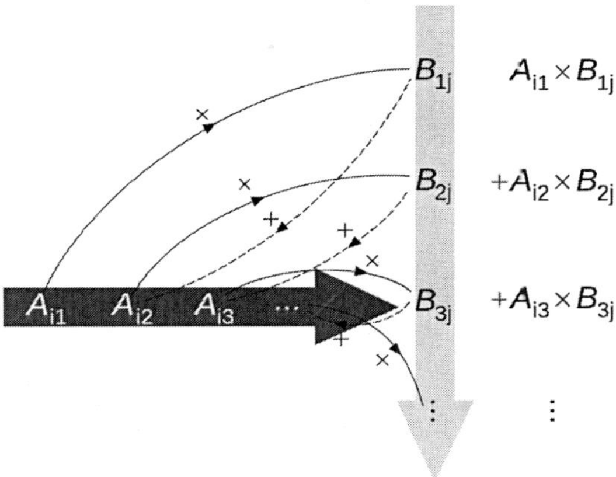

Figture: *The arithmetic process of multiplying numbers in row* i *in matrix A and column* j *in matrix B, then adding to obtain entry* ij *in the final matrix.*

When multiplying matrices, the elements of the rows in the first matrix are multiplied with corresponding columns in the second matrix (depicted in the image right). One may compute each entry in the third matrix one at a time.

For two matrices

$$\mathbf{A} = \begin{pmatrix} A_{11} & A_{12} & \cdots & A_{1m} \\ A_{21} & A_{22} & \cdots & A_{2m} \\ \vdots & \vdots & \ddots & \vdots \\ A_{n1} & A_{n2} & \cdots & A_{nm} \end{pmatrix}, \quad \mathbf{B} = \begin{pmatrix} B_{11} & B_{12} & \cdots & B_{1p} \\ B_{21} & B_{22} & \cdots & B_{2p} \\ \vdots & \vdots & \ddots & \vdots \\ B_{m1} & B_{m2} & \cdots & B_{mp} \end{pmatrix}$$

(where *necessarily* the number of columns in A equals the number of rows in B equals m) the matrix product AB is defined by

$$\mathbf{AB} = \begin{pmatrix} (AB)_{11} & (AB)_{12} & \cdots & (AB)_{1p} \\ (AB)_{21} & (AB)_{22} & \cdots & (AB)_{2p} \\ \vdots & \vdots & \ddots & \vdots \\ (AB)_{n1} & (AB)_{n2} & \cdots & (AB)_{np} \end{pmatrix}$$

(with no multiplication signs or dots) where AB has entries defined by

$$(AB)_{ij} = \sum_{k=1}^{m} A_{ik} B_{kj}$$

Treating the rows and columns in each matrix as row and column vectors respectively, this entry is also their vector dot product:

$$\mathbf{a}_i = \begin{pmatrix} A_{i1} & A_{i2} & \cdots & A_{in} \end{pmatrix}, \quad \mathbf{b}_j = \begin{pmatrix} B_{1j} \\ B_{2j} \\ \vdots \\ B_{nj} \end{pmatrix}, \quad (AB)_{ij} = \mathbf{a}_i \cdot \mathbf{b}_j$$

Usually the entries are numbers or expressions, but can even be matrices themselves.

The matrix product can still be calculated the same way.

Illustration

The figure to the right illustrates diagrammatically the product of two matrices A and B, showing how each intersection in the product matrix corresponds to a row of A and a column of B.

$$\overset{4\times2 \text{ matrix}}{\begin{bmatrix} a_{11} & a_{12} \\ \cdot & \cdot \\ a_{31} & a_{32} \\ \cdot & \cdot \end{bmatrix}} \overset{2\times3 \text{ matrix}}{\begin{bmatrix} \cdot & b_{12} & b_{13} \\ \cdot & b_{22} & b_{23} \end{bmatrix}} = \overset{4\times3 \text{ matrix}}{\begin{bmatrix} \cdot & x_{12} & \cdot \\ \cdot & \cdot & \cdot \\ \cdot & \cdot & x_{33} \\ \cdot & \cdot & \cdot \end{bmatrix}}$$

The values at the intersections marked with circles are:

$$\begin{aligned} x_{12} &= (a_{11}, a_{12}) \cdot (b_{12}, b_{22}) &&= a_{11}b_{12} + a_{12}b_{22} \\ x_{33} &= (a_{31}, a_{32}) \cdot (b_{13}, b_{23}) &&= a_{31}b_{13} + a_{32}b_{23}. \end{aligned}$$

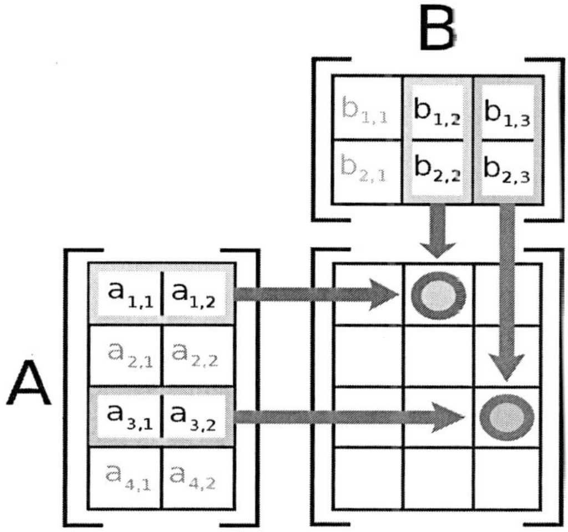

Examples

Square matrix and column vector

$$\mathbf{A} = \begin{pmatrix} a & b \\ c & d \end{pmatrix}, \quad \mathbf{B} = \begin{pmatrix} x \\ y \end{pmatrix}$$

their matrix product is:

$$\mathbf{AB} = \begin{pmatrix} a & b \\ c & d \end{pmatrix}\begin{pmatrix} x \\ y \end{pmatrix} = \begin{pmatrix} a \times x + b \times y \\ c \times x + d \times y \end{pmatrix} = \begin{pmatrix} ax + by \\ cx + dy \end{pmatrix}$$

yet BA is not defined.

The product of a square matrix multiplied by a column matrix arises naturally in linear algebra; for solving linear equations and representing linear transformations.

By choosing *a, b, c, d* in A appropriately, A can represent a variety of transformations such as rotations, scaling and reflections, shears, of a geometric shape in space.

Square matrices

$$\mathbf{A} = \begin{pmatrix} 1 & 2 \\ 3 & 4 \end{pmatrix}, \quad \mathbf{B} = \begin{pmatrix} a & b \\ c & d \end{pmatrix}$$

their matrix products are:

$$\mathbf{AB} = \begin{pmatrix} 1 & 2 \\ 3 & 4 \end{pmatrix}\begin{pmatrix} a & b \\ c & d \end{pmatrix} = \begin{pmatrix} 1\times a+2\times c & 1\times b+2\times d \\ 3\times a+4\times c & 3\times b+4\times d \end{pmatrix}$$

$$= \begin{pmatrix} a+2c & b+2d \\ 3a+4c & 3b+4d \end{pmatrix}$$

and

$$\mathbf{BA} = \begin{pmatrix} a & b \\ c & d \end{pmatrix}\begin{pmatrix} 1 & 2 \\ 3 & 4 \end{pmatrix} = \begin{pmatrix} a\times 1+b\times 3 & a\times 2+b\times 4 \\ c\times 1+d\times 3 & c\times 2+d\times 4 \end{pmatrix}$$

$$= \begin{pmatrix} a+3b & 2a+4b \\ c+3d & 2c+4d \end{pmatrix}$$

Multiplying square matrices which represent linear transformations corresponds to the composite transformation.

Chapter 2

Properties of Matrix Multiplication

General

Analogous to numbers (elements of a field), matrices satisfy the following general properties. Although there is one subtlety, due to the nature of matrix multiplication.

All matrices:

1. *Not commutative (in general):* $\mathbf{AB} \neq \mathbf{BA}$ because AB and BA may not be simultaneously defined, even if they are they may still not be equal. This is contrary to ordinary multiplication of numbers. As long as the entries of the matrix come from a ring which has an identity, and $n > 1$ there is a pair of $n \times n$ noncommuting matrices over the ring.
2. *Associative:* $\mathbf{A(BC)} = \mathbf{(AB)C}$
3. *Distributive over matrix addition:* $\mathbf{A(B+C)} = \mathbf{AB} + \mathbf{AC}$, $\mathbf{(A+B)C} = \mathbf{AC} + \mathbf{BC}$
4. *Scalar multiplication commutes with matrix multiplication:*

   $\lambda(\mathbf{AB}) = (\lambda\mathbf{A})\mathbf{B}$ and $(\mathbf{AB})\lambda = \mathbf{A}(\mathbf{B}\lambda)$ where λ is a scalar. If λ belongs to the centre of the ring of entries of the matrix, then all four quantities are equal, because λX = Xλ for all matrices X. This condition is automatically satisfied if the numbers in the entries come from a commutative ring, for example, a field.
5. *Transpose:* $(\mathbf{AB})^{\mathrm{T}} = \mathbf{B}^{\mathrm{T}}\mathbf{A}^{\mathrm{T}}$ where T denotes the transpose, the interchange of row i with column i in a

matrix. This identity holds for any matrices over a commutative ring, but not for all rings in general.

7. *Hermitian conjugate:* If A and B have complex entries, then $(\mathbf{AB})^\dagger = \mathbf{B}^\dagger \mathbf{A}^\dagger$ where denotes the Hermitian conjugate of a matrix (complex conjugate and transposed).
8. *Traces:* The trace of a product AB is independent of the order of A and B: $\text{tr}(\mathbf{AB}) = \text{tr}(\mathbf{BA})$

Square Matrices Only

1. *Identity element:* If A is a square matrix, then $\mathbf{AI} = \mathbf{IA} = \mathbf{A}$ where I is the identity matrix of the same order.
2. *Inverse matrix:* If A is a square matrix, there *may* be an inverse matrix A^{-1} of A such that $\mathbf{AA}^{-1} = \mathbf{A}^{-1}\mathbf{A} = \mathbf{I}$ If this property holds then A is an invertible matrix, if not A is a singular matrix.
3. *Determinants:* The determinant of a product AB is the product of the determinants of A and B.

While the *products of matrices* AB and BA themselves do not always commute, the *determinants* of the products always do since they are numbers.

In linear algebra, linear transformations can be represented by matrices. If T is a linear transformation mapping $\mathbb{R}^n$ to $\mathbb{R}^m$ and $\vec{x}$ is a column vector with n entries, then

$$T(\vec{x}) = \mathbf{A}\vec{x}$$

for some $m \times n$ matrix A, called the transformation matrix of T. There is an alternative expression of transformation matrices involving row vectors that is preferred by some authors.

Uses

Matrices allow arbitrary linear transformations to be represented in a consistent format, suitable for computation. This also allows transformations to be concatenated easily (by multiplying their matrices).

Linear transformations are not the only ones that can be represented by matrices. Some transformations that are non-linear on a n-dimensional Euclidean space R^n, can be represented as linear transformations on the n+1-dimensional space R^{n+1}.

These include both affine transformations (such as translation) and projective transformations. For this reason, 4x4 transformation matrices are widely used in 3D computer graphics. These n+1-dimensional transformation matrices are called, depending on their application, *affine transformation matrices*, *projective transformation matrices*, or more generally *non-linear transformation matrices*. With respect to an n-dimensional matrix, an n+1-dimensional matrix can be described as an augmented matrix.

In the physical sciences, an active transformation is one which actually changes the physical position of a system, and makes sense even in the absence of a coordinate system whereas a passive transformation is a change in the coordinate description of the physical system (change of basis).

The distinction between active and passive transformations is important. By default, by *transformation*, mathematicians usually mean active transformations, while physicists could mean either.

Put differently, a *passive* transformation refers to observation of the *same* event from two different coordinate frames.

Finding the Matrix of a Transformation

If one has a linear transformation $T(x)$ in functional form, it is easy to determine the transformation matrix A by simply transforming each of the vectors of the standard basis by T and then inserting the results into the columns of a matrix. In other words,

$$\mathbf{A} = \begin{bmatrix} T(\vec{e}_1) & T(\vec{e}_2) & \cdots & T(\vec{e}_n) \end{bmatrix}$$

For example, the function $T(x) = 5x$ is a linear transformation. Applying the above process (suppose that $n = 2$ in this case) reveals that

$$T(\vec{x}) = 5\vec{x} = 5\mathbf{I}\vec{x} = \begin{bmatrix} 5 & 0 \\ 0 & 5 \end{bmatrix}\vec{x}$$

Examples in 2D Graphics

Most common geometric transformations that keep the origin fixed are linear, including rotation, scaling, shearing, reflection, and orthogonal projection; if an affine transformation is not a pure translation it keeps some point fixed, and that point can be chosen as origin to make the transformation linear. In two dimensions, linear transformations can be represented using a 2×2 transformation matrix.

Rotation

For rotation by an angle θ counter clockwise about the origin, the functional form is $x' = x\cos\theta - y\sin\theta$ and $y' = x\sin\theta + y\cos\theta$. Written in matrix form, this becomes:

$$\begin{bmatrix} x' \\ y' \end{bmatrix} = \begin{bmatrix} \cos\theta & -\sin\theta \\ \sin\theta & \cos\theta \end{bmatrix} \begin{bmatrix} x \\ y \end{bmatrix}$$

Similarly, for a rotation clockwise about the origin, the functional form is $x' = x\cos\theta + y\sin\theta$ and $y' = -x\sin\theta + y\cos\theta$ and the matrix form is:

$$\begin{bmatrix} x' \\ y' \end{bmatrix} = \begin{bmatrix} \cos\theta & \sin\theta \\ -\sin\theta & \cos\theta \end{bmatrix} \begin{bmatrix} x \\ y \end{bmatrix}$$

Scaling

For scaling (that is, enlarging or shrinking), we have $x' = s_x \cdot x$ and $y' = s_y \cdot y$ The matrix form is:

$$\begin{bmatrix} x' \\ y' \end{bmatrix} = \begin{bmatrix} s_x & 0 \\ 0 & s_y \end{bmatrix} \begin{bmatrix} x \\ y \end{bmatrix}$$

When $s_x s_y = 1$, then the matrix is a squeeze mapping and preserves areas in the plane.

Shearing

For shear mapping (visually similar to slanting), there are two possibilities. For a shear parallel to the x axis has $x' = x + ky$ and $y' = y$; the shear matrix, applied to column vectors, is:

$$\begin{bmatrix} x' \\ y' \end{bmatrix} = \begin{bmatrix} 1 & k \\ 0 & 1 \end{bmatrix} \begin{bmatrix} x \\ y \end{bmatrix}$$

A shear parallel to the y axis has $x' = x$ and $y' = y + kx$, which has matrix form:

$$\begin{bmatrix} x' \\ y' \end{bmatrix} = \begin{bmatrix} 1 & 0 \\ k & 1 \end{bmatrix} \begin{bmatrix} x \\ y \end{bmatrix}$$

Reflection

To reflect a vector about a line that goes through the origin, let $\vec{l} = (l_x, l_y)$ be a vector in the direction of the line:

$$\mathbf{A} = \frac{1}{\|\vec{l}\|^2} \begin{bmatrix} l_x^2 - l_y^2 & 2l_x l_y \\ 2l_x l_y & l_y^2 - l_x^2 \end{bmatrix}$$

To reflect a point through a plane $ax + by + cz = 0$ (which goes through the origin), one can use $\mathbf{A} = \mathbf{I} - 2\mathbf{N}\mathbf{N}^T$, where I is the 3x3 identity matrix and N is the three-dimensional unit vector for the surface normal of the plane. If the L2 norm of a, b, and c is unity, the transformation matrix can be expressed as:

$$\mathbf{A} = \begin{bmatrix} 1-2a^2 & -2ab & -2ac \\ -2ab & 1-2b^2 & -2bc \\ -2ac & -2bc & 1-2c^2 \end{bmatrix}$$

Note that these are particular cases of a Householder reflection in two and three dimensions. A reflection about a line or plane that does not go through the origin is not a linear transformation; it is an affine transformation.

Orthogonal Projection

To project a vector orthogonally onto a line that goes through the origin, let $\vec{u} = (u_x, u_y)$ be a vector in the direction of the line. Then use the transformation matrix:

$$\mathbf{A} = \frac{1}{\|\vec{u}\|^2} \begin{bmatrix} u_x^2 & u_x u_y \\ u_x u_y & u_y^2 \end{bmatrix}$$

As with reflections, the orthogonal projection onto a line that does not pass through the origin is an affine, not linear, transformation.

Parallel projections are also linear transformations and can be represented simply by a matrix. However, perspective projections are not, and to represent these with a matrix, homogeneous coordinates must be used.

Composing and Inverting Transformations

One of the main motivations for using matrices to represent linear transformations is that transformations can then be easily composed (combined) and inverted. Composition is accomplished by matrix multiplication. If A and B are the matrices of two linear transformations, then the effect of applying first A and then B to a vector x is given by:

$$\mathbf{B}(\mathbf{A}\vec{x}) = (\mathbf{BA})\vec{x}$$

In other words, the matrix of the combined transformation *A followed by B* is simply the product of the individual matrices. Note that the multiplication is done in the opposite order from the English sentence: the matrix of –A followed by B– is BA, not AB.

A consequence of the ability to compose transformations by multiplying their matrices is that transformations can also be inverted by simply inverting their matrices. So, A^{-1} represents the transformation that "undoes" A.

Other Kinds of Transformations

Affine Transformations

To represent affine transformations with matrices, we can use homo-geneous coordinates. This means representing a 2-vector (x, y) as a 3-vector $(x, y, 1)$, and similarly for higher dimensions. Using this system, translation can be expressed with matrix multiplication. The functional form $x' = x + t_x$; $y' = y + t_y$ becomes:

$$\begin{bmatrix} x' \\ y' \\ 1 \end{bmatrix} = \begin{bmatrix} 1 & 0 & t_x \\ 0 & 1 & t_y \\ 0 & 0 & 1 \end{bmatrix} \begin{bmatrix} x \\ y \\ 1 \end{bmatrix}.$$

All ordinary linear transformations are included in the set of affine transformations, and can be described as a simplified form of affine transformations.

Therefore, any linear transformation can be also represented by a general transformation matrix.

The latter is obtained by expanding the corresponding linear transformation matrix by one row and column, filling the extra space with zeros except for the lower-right corner, which must be set to 1. For example, *the anti-clockwise rotation matrix from above* becomes:

$$\begin{bmatrix} \cos\theta & -\sin\theta & 0 \\ \sin\theta & \cos\theta & 0 \\ 0 & 0 & 1 \end{bmatrix}$$

Using transformation matrices containing homogeneous coordinates, translations can be seamlessly intermixed with all other types of transformations.

The reason is that the real plane is mapped to the $w = 1$ plane in real projective space, and so translation in real Euclidean space can be represented as a shear in real projective space. Although a translation is a non-linear transformation in a 2-D or 3-D Euclidean space described by Cartesian coordinates, it becomes, in a 3-D or 4-D projective space described by homogeneous coordinates, a simple linear transformation (a shear).

When using affine transformations, the homogeneous component of a coordinate vector (normally called w) will never be altered. One can therefore safely assume that it is always 1 and ignore it. However, this is not true when using perspective projections.

Perspective Projection

Another type of transformation, of importance in 3D computer graphics, is the perspective projection. Whereas parallel projections are used to project points onto the image plane along parallel lines, the perspective projection projects points onto the image plane along lines that emanate from a single point, called the centre of projection. This means that an object has a smaller projection when it is far away from the centre of projection and a larger projection when it is closer. The simplest perspective projection uses the origin as the centre of projection, and $z = 1$ as the image plane. The functional form of this transformation is then $x' = x / z$; $y' = y / z$. We can express this in homogeneous coordinates as:

$$\begin{bmatrix} x_c \\ y_c \\ z_c \\ w_c \end{bmatrix} = \begin{bmatrix} 1 & 0 & 0 & 0 \\ 0 & 1 & 0 & 0 \\ 0 & 0 & 1 & 0 \\ 0 & 0 & 1 & 0 \end{bmatrix} \begin{bmatrix} x \\ y \\ z \\ 1 \end{bmatrix} = \begin{bmatrix} x \\ y \\ z \\ z \end{bmatrix}$$

After carrying out the matrix multiplication, the homogeneous component w_c will, in general, not be equal to 1. Therefore, to map back into the real plane we must perform the homogeneous divide or perspective divide by dividing each component by w_c:

$$\begin{bmatrix} x' \\ y' \\ z' \end{bmatrix} = \frac{1}{w_c} \begin{bmatrix} x_c \\ y_c \\ z_c \end{bmatrix}$$

More complicated perspective projections can be composed by combining this one with rotations, scales, translations, and shears to move the image plane and centre of projection wherever they are desired.

Examples of Linear Transformation Matrices

In two-dimensional space R^2 linear maps are described by 2×2 real matrices. These are some examples:

- Rotation by 90 degrees counterclockwise: $\mathbf{A} = \begin{pmatrix} 0 & -1 \\ 1 & 0 \end{pmatrix}$
- Rotation by θ degrees counterclockwise: $\mathbf{A} = \begin{pmatrix} \cos(\theta) & -\sin(\theta) \\ \sin(\theta) & \cos(\theta) \end{pmatrix}$
- Reflection against the x axis: $\mathbf{A} = \begin{pmatrix} 1 & 0 \\ 0 & -1 \end{pmatrix}$
- Reflection against the y axis: $\mathbf{A} = \begin{pmatrix} -1 & 0 \\ 0 & 1 \end{pmatrix}$

- Scaling by 2 in all directions: $\mathbf{A} = \begin{pmatrix} 2 & 0 \\ 0 & 2 \end{pmatrix}$
- Horizontal shear mapping: $\mathbf{A} = \begin{pmatrix} 1 & m \\ 0 & 1 \end{pmatrix}$
- Squeeze mapping: $\mathbf{A} = \begin{pmatrix} k & 0 \\ 0 & 1/k \end{pmatrix}$
- Projection onto the y axis: $\mathbf{A} = \begin{pmatrix} 0 & 0 \\ 0 & 1 \end{pmatrix}$.

Forming New Linear Maps from Given Ones

The composition of linear maps is linear: if $f : V \rightarrow W$ and $g : W \rightarrow Z$ are linear, then so is their composition $g \circ f : V \rightarrow Z$. It follows from this that the class of all vector spaces over a given field K, together with K-linear maps as morphisms, forms a category.

The inverse of a linear map, when defined, is again a linear map.

If $f_1 : V \rightarrow W$ and $f_2 : V \rightarrow W$ are linear, then so is their sum $f_1 + f_2$ (which is defined by $(f_1 + f_2)(x) = f_1(x) + f_2(x)$).

If $f : V \rightarrow W$ is linear and a is an element of the ground field K, then the map af, defined by $(af)(x) = a\,(f(x))$, is also linear.

Thus the set $L(V,W)$ of linear maps from V to W itself forms a vector space over K, sometimes denoted $\mathrm{Hom}(V,W)$. Furthermore, in the case that $V=W$, this vector space (denoted $\mathrm{End}(V)$) is an associative algebra under composition of maps, since the composition of two linear maps is again a linear map, and the composition of maps is always associative. This case is discussed in more detail below.

Given again the finite-dimensional case, if bases have been chosen, then the composition of linear maps corresponds to the matrix multiplication, the addition of linear maps corresponds to the matrix addition, and the multiplication of linear maps with scalars corresponds to the multiplication of matrices with scalars.

Endomorphisms and Automorphisms

A linear transformation $f: V \to V$ is an endomorphism of V; the set of all such endomorphisms End(V) together with addition, composition and scalar multiplication as defined above forms an associative algebra with identity element over the field K (and in particular a ring). The multiplicative identity element of this algebra is the identity map id : $V \to V$.

An endomorphism of V that is also an isomorphism is called an automorphism of V.

The composition of two automorphisms is again an automorphism, and the set of all automorphisms of V forms a group, the automorphism group of V which is denoted by Aut(V) or GL(V).

Since the automorphisms are precisely those endomorphisms which possess inverses under composition, Aut(V) is the group of units in the ring End(V).

If V has finite dimension n, then End(V) is isomorphic to the associative algebra of all n by n matrices with entries in K. The automorphism group of V is isomorphic to the general linear group GL(n, K) of all n by n invertible matrices with entries in K.

Kernel, Image and the Rank-Nullity Theorem

If $f: V \to W$ is linear, we define the kernel and the image or range of f by

$$\ker(f) = \{x \in V : f(x) = 0\}$$

$$\text{im}(f) = \{w \in W : w = f(x), x \in V\}$$

ker(f) is a subspace of V and im(f) is a subspace of W. The following dimension formula is known as the rank-nullity theorem:

$$\dim(\ker(f)) + \dim(\text{im}(f)) = \dim(V).$$

The number dim(im(f)) is also called the *rank of f* and written as rank(f), or sometimes, $\rho(f)$; the number dim(ker(f)) is called the *nullity of f* and written as null(f) or $\nu(f)$. If V and W are finite-dimensional, bases have been chosen and f is represented by the matrix A, then the rank and nullity of f are equal to the rank and nullity of the matrix A, respectively.

Cokernel

In mathematics, the cokernel of a linear mapping of vector spaces $f : X \to Y$ is the quotient space $Y/\text{im}(f)$ of the codomain of f by the image of f. The dimension of the cokernel is called the *corank* of f.

Cokernels are dual to the kernels of category theory, hence the name: the kernel is a subobject of the domain (it maps to the domain), while the cokernel is a quotient object of the codomain (it maps from the codomain).

Intuitively, given an equation $f(x) = y$ that one is seeking to solve, the cokernel measures the *constraints* that y must satisfy for this equation to have a solution – the obstructions to a solution – while the kernel measures the *degrees of freedom* in a solution, if one exists. This is elaborated in intuition, below.

More generally, the cokernel of a morphism $f : X \to Y$ in some category (e.g. a homomorphism between groups or a bounded linear operator between Hilbert spaces) is an object Q and a morphism $q : Y \to Q$ such that the composition $q\,f$ is the zero morphism of the category, and furthermore q is universal with respect to this property.

Often the map q is understood, and Q itself is called the cokernel of f. In many situations in abstract algebra, such as for abelian groups, vector spaces or modules, the cokernel of the homomorphism $f : X \to Y$ is the quotient of Y by the image of f. In topological settings, such as with bounded linear operators between Hilbert spaces, one typically has to take the closure of the image before passing to the quotient.

Examples

In the category of groups, the cokernel of a group homomorphism $f : G \to H$ is the quotient of H by the normal closure of the image of f. In the case of abelian groups, since every subgroup is normal, the cokernel is just H modulo the image of f:

$$\text{coker}(f) = H / \text{im}(f).$$

Special Cases

In a preadditive category, it makes sense to add and subtract morphisms. In such a category, the coequalizer of two morphisms f and g (if it exists) is just the cokernel of their difference:

$$coeq(f,g) = coker(g-f)$$

In an abelian category (a special kind of preadditive category) the image and coimage of a morphism f are given by

$$im(f) = ker(cokerf)$$

$$coim(f) = coker(kerf)$$

In particular, every abelian category is normal (and conormal as well). That is, every monomorphism m can be written as the kernel of some morphism. Specifically, m is the kernel of its own cokernel:

$$m = ker(coker(m))$$

Intuition

The cokernel can be thought of as the space of *constraints* that an equation must satisfy, as the space of *obstructions,* just as the kernel is the space of *solutions.*

Formally, one may connect the kernel and the cokernel by the exact sequence

$$0 \to \ker T \to V \to W \to \operatorname{coker} T \to 0.$$

These can be interpreted thus: given a linear equation $T(v) = w$ to solve,

- The kernel is the space of *solutions* to the *homogeneous* equation $T(v) = 0$, and its dimension is the number of *degrees of freedom* in a solution, if it exists;
- The cokernel is the space of *constraints* that must be satisfied if the equation is to have a solution, and its dimension is the number of constraints that must be satisfied for the equation to have a solution.

The dimension of the cokernel plus the dimension of the image (the rank) add up to the dimension of the target space, as the dimension of the quotient space $W/T(V)$ is simply the dimension of the space *minus* the dimension of the image.

As a simple example, consider the map $T:\mathbf{R}^2 \to \mathbf{R}^2$, given by $T(x,y) = (0,y)$. Then for an equation $T(x,y) = (a,b)$ to have a solution, we must have $a = 0$ (one constraint), and in that case the solution space is (x,b), or equivalently stated, $(0,b)+(x,0)$, (one degree of freedom). The kernel may be expressed as the subspace $(x,0) < V$: the value of x is the freedom in a solution – while the cokernel may be expressed via the map $W \to \mathbf{R}^1, (a,b) \mapsto (a)$: given a vector (a,b), the value of a is the *obstruction* to there being a solution.

Index

For a linear operator with finite-dimensional kernel and co-kernel, one may define *index* as:

$$\operatorname{ind} f := \dim \ker f - \dim \operatorname{coker} f,$$

namely the degrees of freedom minus the number of constraints.

For a transformation between finite-dimensional vector spaces, this is just the difference $\dim V - \dim W$, by rank-nullity. This gives an indication of how many solutions or how many constraints one has: if mapping from a larger space to a smaller one, the map may be onto, and thus will have degrees of freedom even without constraints. Conversely, if mapping from a smaller space to a larger one, the map cannot be onto, and thus one will have constraints even without degrees of freedom. The index comes of its own in infinite dimensions: it is how homology is defined, which is a central theory in algebra and algebraic topology; the index of an operator is precisely the Euler characteristic of the 2-term complex $0 \to V \to W \to 0$. In operator theory, the index of Fredholm operators is an object of study, with a major result being the Atiyah–Singer index theorem.

Algebraic Classifications of Linear Transformations

No classification of linear maps could hope to be exhaustive. The following incomplete list enumerates some important classifications that do not require any additional structure on the vector space.

Let V and W denote vector spaces over a field, F. Let $T: V \to W$ be a linear map.

- T is said to be *injective* or a *monomorphism* if any of the following equivalent conditions are true:
  - o T is one-to-one as a map of sets.

- ker$T$ = $\{0_V\}$
- T is monic or left-cancellable, which is to say, for any vector space U and any pair of linear maps $R:U \to V$ and $S:U \to V$, the equation $TR=TS$ implies $R=S$.
- T is left-invertible, which is to say there exists a linear map $S:W \to V$ such that ST is the identity map on V.

* T is said to be *surjective* or an *epimorphism* if any of the following equivalent conditions are true:
  - T is onto as a map of sets.
  - coker T = $\{0_W\}$
  - T is epic or right-cancellable, which is to say, for any vector space U and any pair of linear maps $R:W \to U$ and $S:W \to U$, the equation $RT=ST$ implies $R=S$.
  - T is right-invertible, which is to say there exists a linear map $S:W \to V$ such that TS is the identity map on W.
* T is said to be an *isomorphism* if it is both left- and right-invertible. This is equivalent to T being both one-to-one and onto (a bijection of sets) or also to T being both epic and monic, and so being a bimorphism.
* If T: $V \to V$ is an endomorphism, then:
  - If, for some positive integer n, the n-th iterate of T, T^n, is identically zero, then T is said to be nilpotent.
  - If T T = T, then T is said to be idempotent
  - If $T = k\ I$, where k is some scalar, then T is said to be a scaling transformation or scalar multiplication map.

Change of Basis

Given a linear map whose matrix is A, in the basis B of the space it transforms vectors coordinates [u] as [v]=AÅ–[u]. As vectors change with the inverse of B, its inverse transformation is [v]=BÅ–[v'].

Substituting this in the first expression:

$$B\cdot[v'] = A\cdot B\cdot[u'] \rightarrow [v'] = B^{-1}AB[u']$$

Therefore the matrix in the new basis is $A' = B^{-1}AB$, being B the matrix of the given basis. Therefore linear maps are said to be 1-co 1-contra -variant objects, or type (1, 1) tensors.

Continuity

A *linear transformation* between topological vector spaces, for example normed spaces, may be continuous. If its domain and codomain are the same, it will then be a continuous linear operator. A linear operator on a normed linear space is continuous if and only if it is bounded, for example, when the domain is finite-dimensional. An infinite-dimensional domain may have discontinuous linear operators.

An example of an unbounded, hence discontinuous, linear transformation is differentiation on the space of smooth functions equipped with the supremum norm (a function with small values can have a derivative with large values, while the derivative of 0 is 0). Differentiation is not a continuous operator; its codomain is larger than its domain, because the derivative of a smooth function need not be smooth.

Applications

A specific application of linear maps is for geometric transformations, such as those performed in computer graphics, where the translation, rotation and scaling of 2D or 3D objects is performed by the use of a transformation matrix.

Another application of these transformations is in compiler optimizations of nested-loop code, and in parallelizing compiler techniques.

Matrix Chain Multiplication

Matrix chain multiplication is an optimization problem that can be solved using dynamic programming. Given a sequence of matrices, we want to find the most efficient way to multiply these matrices together. The problem is not actually to *perform* the multiplications, but merely to decide in which order to perform the multiplications. We have many options because matrix multiplication is associative. In other words, no matter how we parenthesize the product, the result will be the same. For example, if we had four matrices *A*, *B*, *C*, and *D*, we would have:

$$(ABC)D = (AB)(CD) = A(BCD) = A(BC)D = ...$$

However, the order in which we parenthesize the product affects the number of simple arithmetic operations needed to compute the product, or the *efficiency*. For example, suppose A is a 10 × 30 matrix, B is a 30 × 5 matrix, and C is a 5 × 60 matrix. Then,

$(AB)C = (10{\times}30{\times}5) + (10{\times}5{\times}60) = 1500 + 3000 = 4500$ operations

$A(BC) = (30{\times}5{\times}60) + (10{\times}30{\times}60) = 9000 + 18000 = 27000$ operations.

Clearly the first method is the more efficient. Now that we have identified the problem, how do we determine the optimal parenthesization of a product of *n* matrices? We could go through each possible paren-thesization (brute force), but this would require time exponential in the number of matrices, which is very slow and impractical for large *n*.

The solution, as we will see, is to break up the problem into a set of related subproblems. By solving subproblems one time and reusing these solutions many times, we can drastically reduce the time required. This is known as dynamic programming.

Hadamard Product (Matrices)

In mathematics, the Hadamard product is a binary operation that takes two matrices of the same dimensions, and produces another matrix where each element *ij* is the product of elements *ij* of the original two matrices. It should not be confused with the more common matrix product. The Hadamard product is associative and distributive, and unlike the matrix product it is also commutative. The Hadamard product is named after French mathematician Jacques Hadamard), and is also known as the entrywise product and the Schur product.

Definition

Formally, for two matrices of the same dimensions

$$A, B \in \mathbb{R}^{m \times n}$$

the Hadamard product $A \circ B$ is a matrix of the same dimensions

$$A \circ B \in \mathbb{R}^{m \times n},$$

with elements given by

$$(A \circ B)_{i,j} = (A)_{i,j} \cdot (B)_{i,j}.$$

Properties

The Hadamard product is commutative, associative and distributive over addition. That is,

$$A \circ B = B \circ A,$$

$$A \circ (B \circ C) = (A \circ B) \circ C,$$

$$A \circ (B + C) = A \circ B + A \circ C.$$

The identity matrix under Hadamard multiplication of two *m-by-n* matrices is *m-by-n* matrix where all elements are equal to 1. This is different from the identity matrix under regular matrix multiplication, where only the elements of the main diagonal are equal to 1. Further-more, a matrix has an inverse under Hadamard multiplication if and only if none of the elements are equal to zero.

For vectors x and y, and corresponding diagonal matrices D_x and D_y with these vectors as their leading diagonals, the following identity holds:

$$x^*(A \circ B)y = \operatorname{tr}(D_x^* A D_y B^T),$$

where x^* denotes the conjugate transpose of . In particular, using vectors of ones, this shows that the sum of all elements in the Hadamard product is the trace of . A related result for square and , is that the row-sums of their Hadamard product are the diagonal elements of

$$\sum_j (A \circ B)_{i,j} = (AB^T)_{i,i}.$$

The Hadamard product is a principal submatrix of the Kronecker product.

Schur Product Theorem

The Hadamard product of two positive-semidefinite matrices is positive-semidefinite. This is known as the Schur product theorem, after German mathematician Issai Schur.

For positive-semidefinite matrices A and B, it is also known that

$$\det(A \circ B) \geq \det(A)\det(B).$$

In Programming Languages

Hadamard multiplication is built into certain programming languages under various names. In MATLAB, it is known as array multiplication, with the symbol .*.

Applications

The Hadamard product appears in lossy compression algorithms such as JPEG.

Transpose

In linear algebra, the transpose of a matrix A is another matrix A^T (also written A2, A^{tr} or A^t) created by any one of the following equivalent actions:

- Reflect A over its main diagonal (which runs top-left to bottom-right) to obtain A^T.
- Write the rows of A as the columns of A^T.
- Write the columns of A as the rows of A^T.

Formally, the ith row, jth column element of A^T is the jth row, ith column element of A:

$$[\mathbf{A}^T]_{ij} = [\mathbf{A}]_{ji}$$

If A is an $m \times n$ matrix then A^T is an $n \times m$ matrix.

Examples

- $\begin{bmatrix} 1 & 2 \end{bmatrix}^T = \begin{bmatrix} 1 \\ 2 \end{bmatrix}.$
- $\begin{bmatrix} 1 & 2 \\ 3 & 4 \end{bmatrix}^T = \begin{bmatrix} 1 & 3 \\ 2 & 4 \end{bmatrix}.$
- $\begin{bmatrix} 1 & 2 \\ 3 & 4 \\ 5 & 6 \end{bmatrix}^T = \begin{bmatrix} 1 & 3 & 5 \\ 2 & 4 & 6 \end{bmatrix}.$

- $\begin{bmatrix} 1 & 2 & 8 \\ 3 & 4 & 3 \\ 5 & 6 & 1 \end{bmatrix}^T = \begin{bmatrix} 1 & 3 & 5 \\ 2 & 4 & 6 \\ 8 & 3 & 1 \end{bmatrix}$.

Properties

For matrices A, B and scalar c we have the following properties of transpose:

1. $(\mathbf{A}^T)^T = \mathbf{A}$Taking the transpose is an involution (self inverse).
2. $(\mathbf{A}+\mathbf{B})^T = \mathbf{A}^T + \mathbf{B}^T$ The transpose respects addition.
3. $(\mathbf{AB})^T = \mathbf{B}^T\mathbf{A}^T$ Note that the order of the factors reverses. From this one can deduce that a square matrix A is invertible if and only if A^T is invertible, and in this case we have $(A^{-1})^T = (A^T)^{-1}$. By induction this result extends to the general case of multiple matrices, where we find that $(A_1A_2...A_{k-1}A_k)^T = A_k^T A_{k-1}^T... A_2^T A_1^T$.
4. $(c\mathbf{A})^T = c\mathbf{A}^T$ The transpose of a scalar is the same scalar. Together with (2), this states that the transpose is a linear map from the space of $m \times n$ matrices to the space of all $n \times m$ matrices.
5. $\det(\mathbf{A}^T) = \det(\mathbf{A})$ The determinant of a square matrix is the same as that of its transpose.
6. The dot product of two column vectors a and b can be computed as $\mathbf{a}\cdot\mathbf{b} = \mathbf{a}^T\mathbf{b}$, which is written as $a_i b^i$ in Einstein notation.
7. If A has only real entries, then A^TA is a positive-semidefinite matrix.
8. $(\mathbf{A}^T)^{-1} = (\mathbf{A}^{-1})^T$ The transpose of an invertible matrix is also invertible, and its inverse is the transpose of the inverse of the original matrix. The notation A^{-T} is often used to represent either of these equivalent expressions.
9. If A is a square matrix, then its eigenvalues are equal to the eigenvalues of its transpose.

Special Transpose Matrices

A square matrix whose transpose is equal to itself is called a symmetric matrix; that is, A is symmetric if

$$\mathbf{A}^{\mathrm{T}} = \mathbf{A}.$$

A square matrix whose transpose is equal to its negative is called skew-symmetric matrix; that is, A is skew-symmetric if

$$\mathbf{A}^{\mathrm{T}} = -\mathbf{A}.$$

The conjugate transpose of the complex matrix A, written as A^*, is obtained by taking the transpose of A and the complex conjugate of each entry:

$$\mathbf{A}^{*} = (\overline{\mathbf{A}})^{\mathrm{T}} = \overline{(\mathbf{A}^{\mathrm{T}})}.$$

A square matrix whose transpose is also its inverse is called an orthogonal matrix; that is, G is orthogonal if $\mathbf{G}\mathbf{G}^{\mathrm{T}} = \mathbf{G}^{\mathrm{T}}\mathbf{G} = \mathbf{I}_n$, the identity matrix, i.e. $\mathbf{G}^{\mathrm{T}} = \mathbf{G}^{"1}$.

Transpose of Linear Maps

If $f: V \to W$ is a linear map between vector spaces V and W with nondegenerate bilinear forms, we define the *transpose* of f to be the linear map ${}^{t}f: W \to V$, determined by

$$B_V(v,{}^{t}f(w)) = B_W(f(v), w) \quad \forall\, v \in V, w \in W.$$

Here, B_V and B_W are the bilinear forms on V and W respectively. The matrix of the transpose of a map is the transposed matrix only if the bases are orthonormal with respect to their bilinear forms.

Over a complex vector space, one often works with sesquilinear forms instead of bilinear (conjugate-linear in one argument).

The transpose of a map between such spaces is defined similarly, and the matrix of the transpose map is given by the conjugate transpose matrix if the bases are orthonormal. In this case, the transpose is also called the Hermitian adjoint.

If V and W do not have bilinear forms, then the transpose of a linear map $f: V \to W$ is only defined as a linear map ${}^{t}f: W^* \to V^*$ between the dual spaces of W and V.

This means that the transpose (and even the orthogonal group) can be defined abstractly, and completely without reference to matrices (nor the components thereof).

If $f : V \to W$ then for any $o : W \to F$ (that is, any o belonging to W*), if ${}^{T}f(o)$ is defined as o composed with f then it will map $V \to F$ (that is, ${}^{T}f$ will map W* to V*).

If the vector spaces have metrics then V* can be uniquely mapped to V, etc., such that we can immediately consider whether or not $f^{T} : W \to V$ is equal to $f^{-1} : W \to V$.

Implementation of Matrix Transposition on Computers

On a computer, one can often avoid explicitly transposing a matrix in memory by simply accessing the same data in a different order. For example, software libraries for linear algebra, such as BLAS, typically provide options to specify that certain matrices are to be interpreted in transposed order to avoid the necessity of data movement.

However, there remain a number of circumstances in which it is necessary or desirable to physically reorder a matrix in memory to its transposed ordering. For example, with a matrix stored in row-major order, the rows of the matrix are contiguous in memory and the columns are discontiguous. If repeated operations need to be performed on the columns, for example in a fast Fourier transform algorithm, transposing the matrix in memory (to make the columns contiguous) may improve performance by increasing memory locality.

Ideally, one might hope to transpose a matrix with minimal additional storage. This leads to the problem of transposing an $n \times m$ matrix in-place, with O(1) additional storage or at most storage much less than mn. For $n \neq m$, this involves a complicated permutation of the data elements that is non-trivial to implement in-place. Therefore efficient in-place matrix transposition has been the subject of numerous research publications in computer science, starting in the late 1950s, and several algorithms have been developed.

Chapter 3

Abstract Algebraic Aspects

Determinant of An Endomorphism

The above identities concerning the determinant of a products and inverses of matrices imply that similar matrices have the same determinant: two matrices A and B are similar, if there exists an invertible matrix X such that $A = X^{-1}BX$. Indeed, repeatedly applying the above identities yields

$$\det(A) = \det(X)^{-1}\det(BX) = \det(X)^{-1}\det(B)\det(X) =$$
$$\det(B)\det(X)^{-1}\det(X) = \det(B).$$

The determinant is therefore also called a similarity invariant. The determinant of a linear transformation

$$T : V \to V$$

for some finite dimensional vector space V is defined to be the determinant of the matrix describing it, with respect to an arbitrary choice of basis in V. By the similarity invariance, this determinant is independent of the choice of the basis for V and therefore only depends on the endomorphism T.

Exterior Algebra

The determinant can also be characterized as the unique function

$$D : M_n(K) \to K$$

from the set of all n-by-n matrices with entries in a field K to this field satisfying the following three properties: first, D is an n-linear function: considering all but one column of A fixed, the determinant is linear in the remaining column, that is

$$D(v_1,\ldots,v_{i-1},av_i+bw,v_{i+1},\ldots,v_n)=aD(v_1,\ldots,v_{i-1},v_i,v_{i+1},\ldots,v_n)$$
$$+bD(v_1,\ldots,v_{i-1},w,v_{i+1},\ldots,v_n)$$

for any column vectors v_1, ..., v_n, and w and any scalars (elements of K) a and b. Second, D is an alternating function: for any matrix A with two identical columns $D(A) = 0$. Finally, $D(I_n) = 1$. Here I_n is the identity matrix.

This fact also implies that every other n-linear alternating function F: $M_n(K) \rightarrow K$ satisfies

$$F(M) = F(I)D(M).$$

The last part in fact follows from the preceding statement: one easily sees that if F is nonzero it satisfies $F(I) \neq 0$, and function that associates $F(M)/F(I)$ to M satisfies all conditions of the theorem. The importance of stating this part is mainly that it remains valid if K is any commutative ring rather than a field, in which case the given argument does not apply.

The determinant of a linear transformation $A : V \rightarrow V$ of an n-dimensional vector space V can be formulated in a coordinate-free manner by considering the n-th exterior power $\wedge^n V$ of V. A induces a linear map

$$\Lambda^n A : \Lambda^n V \rightarrow \Lambda^n V$$
$$v_1 \wedge v_2 \wedge \ldots \wedge v_n \mapsto Av_1 \wedge Av_2 \wedge \ldots \wedge Av_n.$$

As $\wedge^n V$ is one-dimensional, the map $\wedge^n$A is given by multiplying with some scalar. This scalar coincides with the determinant of A, that is to say

$$(\Lambda^n A)(v_1 \wedge \ldots \wedge v_n) = \det(A)\cdot v_1 \wedge \ldots \wedge v_n.$$

This definition agrees with the more concrete coordinate-dependent definition. This follows from the characterization of the determinant given above. For example, switching two columns changes the parity of the determinant; likewise, permuting the vectors in the exterior product $v_1 \neq v_2 \neq \ldots \neq v_n$ to $v_2 \neq v_1 \neq v_3 \neq \ldots \neq v_n$, say, also alters the parity.

For this reason, the highest non-zero exterior power $\wedge^n(V)$ is sometimes also called the determinant of V and similarly for more involved objects such as vector bundles or chain complexes of vector spaces. Minors of a matrix can also be cast in this setting, by considering lower alternating forms $\wedge^k V$ with k < n.

Square matrices over commutative rings and abstract properties

The determinant of a matrix can be defined, for example using the Leibniz formula, for matrices with entries in any commutative ring.

Briefly, a ring is a structure where addition, subtraction, and multiplication are defined. The commutativity requirement means that the product does not depend on the order of the two factors, i.e.,

$$r \cdot s = s \cdot r$$

is supposed to hold for all elements r and s of the ring. For example, the integers form a commutative ring.

Many of the above statements and notions carry over mutatis mutandis to determinants of these more general matrices: the determinant is multiplicative in this more general situation, and Cramer's rule also holds.

A square matrix over a commutative ring R is invertible if and only if its determinant is a unit in R, that is, an element having a (multiplicative) inverse. (If R is a field, this latter condition is equivalent to the determinant being nonzero, thus giving back the above characterization.) For example, a matrix A with entries in Z, the integers, is invertible (in the sense that the inverse matrix has again integer entries) if the determinant is +1 or −1. Such a matrix is called unimodular.

The determinant defines a mapping

$$\mathrm{GL}_n(R) \rightarrow R^{\times},$$

between the group of invertible $n \times n$ matrices with entries in R and the multiplicative group of units in R. Since it respects the multiplication in both groups, this map is a group homomorphism. Secondly, given a ring homomorphism f: $R \rightarrow S$, there is a map $\mathrm{GL}_n(R) \rightarrow \mathrm{GL}_n(S)$ given by replacing all entries in R by their images under f. The determinant respects these maps, i.e., given a matrix $A = (a_{i,j})$ with entries in R, the identity

$$f(\det((a_{i,j}))) = \det((f(a_{i,j})))$$

holds. For example, the determinant of the complex conjugate of a complex matrix (which is also the determinant of its conjugate transpose) is the complex conjugate of its determinant, and for integer matrices: the reduction modulo m of the determinant of such a matrix is equal to the determinant of the matrix reduced modulo m (the latter determinant being computed using modular arithmetic). In the more high-brow parlance of category theory, the determinant is a natural transformation between the two functors GL_n and $(Å–)^{\times}$. Adding yet another layer of abstraction, this is captured by saying that the determinant is a morphism of algebraic groups, from the general linear group to the multiplicative group,

$$\det : GL_n \rightarrow \mathbb{G}_m.$$

Generalizations and Related Notions

Infinite Matrices

For matrices with an infinite number of rows and columns, the above definitions of the determinant do not carry over directly. For example, in Leibniz' formula, an infinite sum (all of whose terms are infinite products) would have to be calculated. Functional analysis provides different extensions of the determinant for such infinite-dimensional situations, which however only work for particular kinds of operators.

The Fredholm determinant defines the determinant for operators known as trace class operators by an appropriate generalization of the formula

$$\det(I + A) = \exp(\mathrm{tr}(\log(I + A))).$$

Another infinite-dimensional notion of determinant is the functional determinant.

Notions of Determinant Over Non-Commutative Rings

For square matrices with entries in a non-commutative ring, there are various difficulties in defining determinants in a manner analogous to that for commutative rings.

A meaning can be given to the Leibniz formula provided the order for the product is specified, and similarly for other ways to define the determinant, but non-commutativity then leads to the loss of many fundamental properties of the determinant, for instance the multiplicative property or the fact that the determinant is unchanged under transposition of the matrix.

Over non-commutative rings, there is no reasonable notion of a multilinear form (if a bilinear form exists with a regular element of R as value on some pair of arguments, it can be used to show that all elements of R commute). Nevertheless various notions of non-commutative determinant have been formulated, which preserve some of the properties of determinants, notably quasideterminants and the Dieudonné determinant.

Further Variants

Determinants of matrices in superrings (that is, Z/2-graded rings) are known as Berezinians or superdeterminants. The permanent of a matrix is defined as the determinant, except that the factors sgn(σ) occurring in Leibniz' rule are omitted. The immanant generalizes both by introducing a character of the symmetric group S_n in Leibniz' rule.

Calculation

Determinants are mainly used as a theoretical tool. They are rarely calculated explicitly in numerical linear algebra, where for applications like checking invertibility and finding eigenvalues the determinant has largely been supplanted by other techniques. Nonetheless, explicitly calculating determinants is required in some situations, and different methods are available to do so.

Naive methods of implementing an algorithm to compute the determinant include using Leibniz' formula or Laplace's formula. Both these approaches are extremely inefficient for large matrices, though, since the number of required operations grows very quickly: it is of order $n!$ (n factorial) for an $n \times n$ matrix M. For example, Leibniz' formula requires to calculate $n!$ products. Therefore, more involved techniques have been developed for calculating determinants.

Decomposition Methods

Given a matrix A, some methods compute its determinant by writing A as a product of matrices whose determinants can be more easily computed. Such techniques are referred to as decomposition methods. Examples include the LU decomposition, the QR decomposition or the Cholesky decomposition (for positive definite matrices). These methods are of order $O(n^3)$, which is a significant improvement over $O(n!)$

The LU decomposition expresses A in terms of a lower triangular matrix L, an upper triangular matrix U and a permutation matrix P:

$$A = PLU.$$

The determinants of L and U can be quickly calculated, since they are the products of the respective diagonal entries. The determinant of P is just the sign ε of the corresponding permutation (which is +1 for an even number of permutations and is -1 for an uneven number of permutations). The determinant of A is then

$$\det(A) = \varepsilon \det(L)\cdot\det(U),$$

Moreover, the decomposition can be chosen such that L is a unitriangular matrix and therefore has determinant 1, in which case the formula further simplifies to

$$\det(A) = \varepsilon \det(U).$$

Applications

Linear Independence

As mentioned above, the determinant of a matrix (with real or complex entries, say) is zero if and only if the column vectors of the matrix are linearly dependent. Thus, determinants can be used to characterize linearly dependent vectors. For example, given two vectors v_1, v_2 in R^3, a third vector v_3 lies in the plane spanned by the former two vectors exactly if the determinant of the 3-by-3 matrix consisting of the three vectors is zero. The same idea is also used in the theory of differential equations: given n functions $f_1(x)$, ..., $f_n(x)$ (supposed to be n–1 times differentiable), the Wronskian is defined to be

$$W(f_1,\ldots,f_n)(x) = \begin{vmatrix} f_1(x) & f_2(x) & \cdots & f_n(x) \\ f_{1'}(x) & f_{2'}(x) & \cdots & f_{n'}(x) \\ \vdots & \vdots & \ddots & \vdots \\ f_1^{(n-1)}(x) & f_2^{(n-1)}(x) & \cdots & f_n^{(n-1)}(x) \end{vmatrix}.$$

It is non-zero (for some x) in a specified interval if and only if the given functions and all their derivatives up to order n–1 are linearly independent. If it can be shown that the Wronskian is zero everywhere on an interval then, in the case of analytic functions, this implies the given functions are linearly dependent.

Orientation of a Basis

The determinant can be thought of as assigning a number to every sequence of n in R^n, by using the square matrix whose columns are the given vectors. For instance, an orthogonal matrix with entries in R^n represents an orthonormal basis in Euclidean space. The determinant of such a matrix determines whether the orientation of the basis is consistent with or opposite to the orientation of the standard basis. Namely, if the determinant is +1, the basis has the same orientation. If it is −1, the basis has the opposite orientation.

More generally, if the determinant of A is positive, A represents an orientation-preserving linear transformation (if A is an orthogonal 2×2 or 3×3 matrix, this is a rotation), while if it is negative, A switches the orientation of the basis.

Circulant Matrix

In linear algebra, a circulant matrix is a special kind of Toeplitz matrix where each row vector is rotated one element to the right relative to the preceding row vector. In numerical analysis, circulant matrices are important because they are diagonalized by a discrete Fourier transform, and hence linear equations that contain them may be quickly solved using a fast Fourier transform. They can be interpreted analytically as the integral kernel of a convolution operator on the cyclic group $\mathbf{Z}/n\mathbf{Z}$. In cryptography, a circulant matrix is used in the MixColumns step of the Advanced Encryption Standard.

Definition

An $n \times n$ circulant matrix C takes the form

$$C = \begin{bmatrix} c_0 & c_{n-1} & \cdots & c_2 & c_1 \\ c_1 & c_0 & c_{n-1} & & c_2 \\ \vdots & c_1 & c_0 & \ddots & \vdots \\ c_{n-2} & & \ddots & \ddots & c_{n-1} \\ c_{n-1} & c_{n-2} & \cdots & c_1 & c_0 \end{bmatrix}.$$

A circulant matrix is fully specified by one vector, C, which appears as the first column of C. The remaining columns of C

are each *cyclic permutations* of the vector *C* with offset equal to the column index. The last row of is the vector in reverse order, and the remaining rows are each cyclic permutations of the last row. Note that different sources define the circulant matrix in different ways, for example with the coefficients corresponding to the first row rather than the first column of the matrix, or with a different direction of shift.

Application in Graph Theory

In graph theory, a graph or digraph whose adjacency matrix is circulant is called a circulant graph (or digraph). Equivalently, a graph is circulant if its automorphism group contains a full-length cycle. The Möbius ladders are examples of circulant graphs, as are the Paley graphs for fields of prime order.

Symmetry

A square matrix A that is equal to its transpose, i.e., $A = A^T$, is a symmetric matrix. If instead, A was equal to the negative of its transpose, i.e., $A = -A^T$, then A is a skew-symmetric matrix. In complex matrices, symmetry is often replaced by the concept of Hermitian matrices, which satisfy $A^- = A$, where the star or asterisk denotes the conjugate transpose of the matrix, i.e., the transpose of the complex conjugate of A.

By the spectral theorem, real symmetric matrices and complex Hermitian matrices have an eigenbasis; i.e., every vector is expressible as a linear combination of eigenvectors. In both cases, all eigenvalues are real. This theorem can be generalized to infinite-dimensional situations related to matrices with infinitely many rows and columns.

Definiteness

Computational Aspects

In addition to theoretical knowledge of properties of matrices and their relation to other fields, it is important for practical purposes to perform matrix calculations effectively and precisely. The domain studying these matters is called numerical linear algebra. As with other numerical situations, two main aspects are the complexity of algorithms and their numerical stability. Many problems can be solved by both direct algorithms or iterative approaches. For example, finding eigenvectors can be done by finding a sequence of vectors x_n converging to an eigenvector when n tends to infinity.

Determining the complexity of an algorithm means finding upper bounds or estimates of how many elementary operations such as additions and multiplications of scalars are necessary to perform some algorithm, e.g., multiplication of matrices.

For example, calculating the matrix product of two n-by-n matrix using the definition given above needs n^3 multiplications, since for any of the n^2 entries of the product, n multiplications are necessary.

The Strassen algorithm outperforms this "naive" algorithm; it needs only $n^{2.807}$ multiplications. A refined approach also incorporates specific features of the computing devices.

In many practical situations additional information about the matrices involved is known. An important case are sparse matrices, i.e., matrices most of whose entries are zero.

There are specifically adapted algorithms for, say, solving linear systems Ax = b for sparse matrices A, such as the conjugate gradient method. An algorithm is, roughly speaking, numerically stable, if little deviations (such as rounding errors) do not lead to big deviations in the result.

For example, calculating the inverse of a matrix via Laplace's formula (Adj (A) denotes the adjugate matrix of A)

$$A^{-1} = \mathrm{Adj}(A) / \det(A)$$

may lead to significant rounding errors if the determinant of the matrix is very small. The norm of a matrix can be used to capture the conditioning of linear algebraic problems, such as computing a matrix' inverse.[32]

Although most computer languages are not designed with commands or libraries for matrices, as early as the 1970s, some engineering desktop computers such as the HP 9830 had ROM cartridges to add BASIC commands for matrices.

Some computer languages such as APL were designed to manipulate matrices, and various mathematical programs can be used to aid computing with matrices.

Matrix Decomposition

In the mathematical discipline of linear algebra, a matrix decomposition is a factorization of a matrix into some canonical form. There are many different matrix decompositions; each finds use among a particular class of problems.

Example

In numerical analysis, different decompositions are used to implement efficient matrix algorithms. For instance, when solving a system of linear equations $Ax = b$, the matrix A can be decomposed via the LU decomposition. The LU decomposition factorizes a matrix into a lower triangular matrix L and an upper triangular matrix U. The systems $L(Ux) = b$ and require fewer additions and multiplications to solve, compared with the original system $Ax = b$, though one might require significantly more digits in inexact arithmetic such as floating point.

Similarly, the QR decomposition expresses A as QR with Q a unitary matrix and R an upper triangular matrix. The system $Q(Rx) = b$ is solved by $Rx = Q^{T}b = c$, and the system $Rx = c$ is solved by 'back substitution'. The number of additions and multiplications required is about twice that of using the LU solver, but no more digits are required in inexact arithmetic because the QR decomposition is numerically stable.

Decompositions Related to Solving Systems of Linear Equations

LU Decomposition

- *Applicable to:* square matrix A
- *Decomposition:* $A = LU$, where L is lower triangular and U is upper triangular
- *Related:* the LDU decomposition is $A = LDU$, where L is lower triangular with ones on the diagonal, U is upper triangular with ones on the diagonal, and D is a diagonal matrix.
- *Related:* the LUP decomposition is $A = LUP$, where L is lower triangular, U is upper triangular, and P is a permutation matrix.

- *Existence:* An LUP decomposition exists for any square matrix *A*. When *P* is an identity matrix, the LUP decomposition reduces to the LU decomposition. If the LU decomposition exists, the LDU decomposition does too.
- *Comments:* The LUP and LU decompositions are useful in solving an *n*-by-*n* system of linear equations $Ax = b$. These decompositions summarize the process of Gaussian elimination in matrix form.

  Matrix *P* represents any row interchanges carried out in the process of Gaussian elimination. If Gaussian elimination produces the row echelon form without requiring any row interchanges, then *P=I*, so an LU decomposition exists.

Cholesky decomposition

- *Applicable to:* square, symmetric, positive definite matrix *A*
- *Decomposition:* $A = U^T U$, where *U* is upper triangular with positive diagonal entries
- *Comment:* the Cholesky decomposition is a special case of the symmetric LU decomposition, with $L = U^T$.
- *Comment:* the Cholesky decomposition is unique
- *Comment:* the Cholesky decomposition is also applicable for complex hermitian positive definite matrices
- *Comment:* An alternative is the LDL decomposition which can avoid extracting square roots.

QR decomposition

- *Applicable to:* *m*-by-*n* matrix *A*
- *Decomposition:* $A = QR$ where *Q* is an orthogonal matrix of size *m*-by-*m*, and *R* is an upper triangular matrix of size *m*-by-*n*
- *Comment:* The QR decomposition provides an alternative way of solving the system of equations $Ax = b$ without inverting the matrix *A*. The fact that *Q* is orthogonal means that $Q^T Q = I$, so that $Ax = b$ is equivalent to $Rx = Q^T b$, which is easier to solve since *R* is triangular.

RRQR factorization

Singular value decomposition

- *Applicable to:* m-by-n matrix A.
- *Decomposition:* $A = UDV^H$, where D is a nonnegative diagonal matrix, and U and V are unitary matrices, and V^H denotes the conjugate transpose of V (or simply the transpose, if V contains real numbers only).
- *Comment:* The diagonal elements of D are called the singular values of A.
- *Comment:* like the eigendecomposition below, the singular value decomposition involves finding basis directions along which matrix multiplication is equivalent to scalar multiplication, but it has greater generality since the matrix under consideration need not be square.

Decompositions Based on Eigenvalues and Related Concepts

Eigendecomposition

- Also called *spectral decomposition*
- *Applicable to:* square matrix A with distinct eigen values.
- *Decomposition:* $A = VDV^{-1}$, where D is a diagonal matrix formed from the eigenvalues of A, and the columns of V are the corresponding eigenvectors of A.
- *Existence:* An n-by-n matrix A always has n eigenvalues, which can be ordered (in more than one way) to form an n-by-n diagonal matrix D and a corresponding matrix of nonzero columns V that satisfies the eigenvalue equation $AV = VD$. If the n eigenvalues are distinct (that is, none is equal to any of the others), then V is invertible, implying the decomposition $A = VDV^{-1}$.
- *Comment:* The condition of having n distinct eigenvalues is sufficient but not necessary. The necessary and sufficient condition is for each eigenvalue to have geometric multiplicity equal to its algebraic multiplicity.
- *Comment:* The eigendecomposition is useful for understanding the solution of a system of linear ordinary

differential equations or linear difference equations. For example, the difference equation $x_{t+1} = Ax_t$ starting from the initial condition $x_0 = c$ is solved by , which is equivalent to $x_t = A^t c$, where V and D are the matrices formed from the eigenvectors and eigenvalues of A. Since D is diagonal, raising it to power , just involves raising each element on the diagonal to the power t. This is much easier to do and to understand than raising A to power t, since A is usually not diagonal.

Jordan Decomposition

The Jordan normal form and the Jordan–Chevalley decomposition

- *Applicable to:* square matrix A
- *Comment:* the Jordan normal form generalizes the eigende-composition to cases where there are repeated eigenvalues and cannot be diagonalized, the Jordan–Chevalley decomposition does this without choosing a basis.

Schur decomposition

- *Applicable to:* square matrix A
- *Comment:* there are two versions of this decomposition: the complex Schur decomposition and the real Schur decomposition. A complex matrix always has a complex Schur decomposition. A real matrix admits a real Schur decomposition if and only if all of its eigenvalues are real.
- *Decomposition (complex version):* $A = UTU^H$, where U is a unitary matrix, U^H is the conjugate transpose of U, and T is an upper triangular matrix called the complex Schur form which has the eigenvalues of A along its diagonal.
- *Decomposition (real version):* $A = VSV^T$, where A, V, S and V_T are matrices that contain real numbers only. In this case, V is an orthogonal matrix, is the transpose of V, and S is a block upper triangular matrix called the real Schur form. The blocks on the diagonal of S are of size 1×1 (in which case they represent real

eigenvalues) or 2×2 (in which case they are derived from complex conjugate eigenvalue pairs).

QZ decomposition

- *Also called: generalized Schur decomposition*
- *Applicable to:* square matrices A and B
- *Comment:* there are two versions of this decomposition: complex and real.
- *Decomposition (complex version):* $A = QSZ^H$ and $B = QTZ^H$ where Q and Z are unitary matrices, the H superscript represents conjugate transpose, and S and T are upper triangular matrices.
- *Comment:* in the complex QZ decomposition, the ratios of the diagonal elements of S to the corresponding diagonal elements of T, $\lambda_i = S_{ii} / T_{ii}$, are the generalized eigenvalues that solve the generalized eigenvalue problem $Av = \lambda Bv$ (where is an unknown scalar and v is an unknown nonzero vector).
- *Decomposition (real version):* $A = QSZ^T$ and $B = QTZ^T$ where A, B, Q, Z, S, and T are matrices containing real numbers only. In this case Q and Z are orthogonal matrices, the T superscript represents transposition, and S and T are block upper triangular matrices. The blocks on the diagonal of S and T are of size 1×1 or 2×2.

Chapter 4

Linear Equation

A linear equation is an algebraic equation in which each term is either a constant or the product of a constant and (the first power of) a single variable. Linear equations can have one or more variables. Linear equations occur with great regularity in applied mathematics.

While they arise quite naturally when modeling many phenomena, they are particularly useful since many non-linear equations may be reduced to linear equations by assuming that quantities of interest vary to only a small extent from some "background" state. Linear equations do not include exponents.

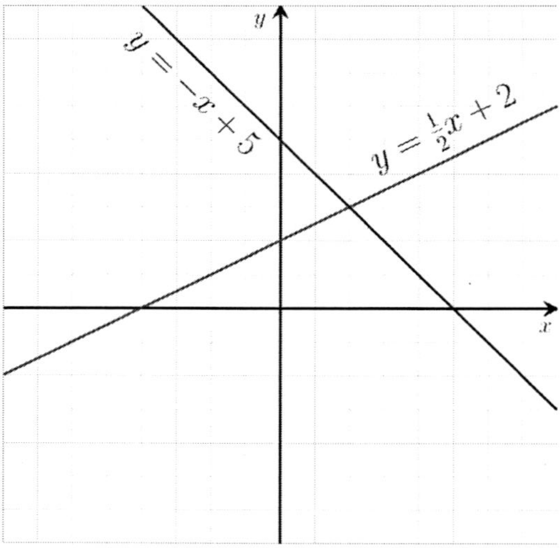

Figure: *Graph sample of linear equations.*

Linear Equations in Two Variables

A common form of a linear equation in the two variables x and y is

$$y = mx + b,$$

where m and b designate constants. The origin of the name "linear" comes from the fact that the set of solutions of such an equation forms a straight line in the plane. In this particular equation, the constant m determines the slope or gradient of that line, and the constant term "b" determines the point at which the line crosses the y-axis, otherwise known as the y-intercept.

Since terms of linear equations cannot contain products of distinct or equal variables, nor any power (other than 1) or other function of a variable, equations involving terms such as xy, x^2, $y^{1/3}$, and $\sin(x)$ are *nonlinear*.

Forms for 2D Linear Equations

Linear equations can be rewritten using the laws of elementary algebra into several different forms. These equations are often referred to as the "equations of the straight line." In what follows, x, y, t, and θ are variables; other letters represent constants (fixed numbers).

General form: $Ax + By + C = 0$, where A and B are not both equal to zero. The equation is usually written so that $A \geq 0$, by convention. The graph of the equation is a straight line, and every straight line can be represented by an equation in the above form. If A is nonzero, then the x-intercept, that is, the x-coordinate of the point where the graph crosses the x-axis (where, y is zero), is "C/A. If B is nonzero, then the y-intercept, that is the y-coordinate of the point where the graph crosses the y-axis (where x is zero), is $-C/B$, and the slope of the line is $-A/B$.

Standard form: $Ax + By = C$, where A and B are not both equal to zero, A, B, and C are coprime integers, and A is nonnegative (if zero, B must be positive). The standard form can be converted to the general form, but not always to all the other forms if A or B is zero. It is worth noting that, while the term occurs frequently in school-level US algebra textbooks, most lines cannot be described by such equations. For instance, the line $x + y = -2$ cannot be described by a linear equation with integer coefficients since -2 is irrational.

Matrix form: Using the order of the standard form $Ax + By = C,$ but without the restriction of coprime integer coefficients one can rewrite the equation in matrix form:

$$\begin{pmatrix} A & B \end{pmatrix}\begin{pmatrix} x \\ y \end{pmatrix} = \begin{pmatrix} C \end{pmatrix}.$$

Further, this representation extends to systems of linear equations.

$$A_1x + B_1y = C_1,$$

$$A_2x + B_2y = C_2,$$

becomes

$$\begin{pmatrix} A_1 & B_1 \\ A_2 & B_2 \end{pmatrix}\begin{pmatrix} x \\ y \end{pmatrix} = \begin{pmatrix} C_1 \\ C_2 \end{pmatrix}.$$

Since this extends easily to higher dimensions, it is a usual method in linear algebra, and in computer programming. In particular, there are named methods for solving simultaneous linear equations like Gauss-Jordan which can be expressed as matrix elementary row operations.

Linear Equations in More than Two Variables

A linear equation can involve more than two variables. The general linear equation in n variables is:

$$a_1x_1 + a_2x_2 + \cdots + a_nx_n = b.$$

In this form, a_1, a_2, ..., a_n are the coefficients, x_1, x_2, ..., x_n are the variables, and b is the constant. When dealing with three or fewer variables, it is common to replace x_1 with just x, x_2 with y, and x_3 with z, as appropriate.

Such an equation will represent an $(n-1)$-dimensional hyperplane in n-dimensional Euclidean space (for example, a plane in 3-space).

In vector notation, this can be expressed as:

$$\vec{n} \cdot \vec{x} = \vec{n} \cdot \overrightarrow{x_0}$$

where $\vec{n}$ is a vector normal to the plane, $\vec{x}$ are the coordinates of any point on the plane, and $\overrightarrow{x_0}$ are the coordinates of the origin of the plane.

System of Linear Equations

In mathematics, a system of linear equations (or linear system) is a collection of linear equations involving the same set of variables. For example,

$$\begin{array}{rcrcrcr} 3x & + & 2y & - & z & = & 1 \\ 2x & - & 2y & + & 4z & = & -2 \\ -x & + & \frac{1}{2}y & - & z & = & 0 \end{array}$$

is a system of three equations in the three variables x, y, z. A solution to a linear system is an assignment of numbers to the variables such that all the equations are simultaneously satisfied. A solution to the system above is given by

$$\begin{aligned} x &= 1 \\ y &= -2 \\ z &= -2 \end{aligned}$$

since it makes all three equations valid.

In mathematics, the theory of linear systems is the basis and a fundamental part of linear algebra, a subject which is used in most parts of modern mathematics.

Computational algorithms for finding the solutions are an important part of numerical linear algebra, and play a prominent role in engineering, physics, chemistry, computer science, and economics.

A system of non-linear equations can often be approximated by a linear system, a helpful technique when making a mathematical model or computer simulation of a relatively complex system.

Very often, the coefficients of the equations are real or complex numbers and the solutions are searched in the same set of numbers, but the theory and the algorithms apply for coefficients and solutions in any field.

For solutions in an integral domain like the ring of the integers, or in other algebraic structures, other theories have been developed.

Elementary Example

The simplest kind of linear system involves two equations and two variables:

$$\begin{array}{ccccc} 2x & + & 3y & = & 6 \\ 4x & + & 9y & = & 15 \end{array}.$$

One method for solving such a system is as follows. First, solve the top equation for x in terms of y:

$$x = 3 - \frac{3}{2}y.$$

Now substitute this expression for x into the bottom equation:

$$4\left(3 - \frac{3}{2}y\right) + 9y = 15.$$

This results in a single equation involving only the variable y. Solving gives $y = 1$, and substituting this back into the equation for x yields $x = 3/2$. This method generalizes to systems with additional variables.

General Form

A general system of m linear equations with n unknowns can be written as

$$\begin{array}{ccccccc} a_{11}x_1 & + & a_{12}x_2 & + \cdots + & a_{1n}x_n & = & b_1 \\ a_{21}x_1 & + & a_{22}x_2 & + \cdots + & a_{2n}x_n & = & b_2 \\ \vdots & & \vdots & & \vdots & & \vdots \\ a_{m1}x_1 & + & a_{m2}x_2 & + \cdots + & a_{mn}x_n & = & b_m. \end{array}$$

Here $x_1, x_2, \ldots, x_n$ are the unknowns, $a_{11}, a_{12}, \ldots, a_{mn}$ are the coefficients of the system, and $b_1, b_2, \ldots, b_m$ are the constant terms.

Often the coefficients and unknowns are real or complex numbers, but integers and rational numbers are also seen, as are polynomials and elements of an abstract algebraic structure.

Vector Equation

One extremely helpful view is that each unknown is a weight for a column vector in a linear combination.

$$x_1 \begin{bmatrix} a_{11} \\ a_{21} \\ \vdots \\ a_{m1} \end{bmatrix} + x_2 \begin{bmatrix} a_{12} \\ a_{22} \\ \vdots \\ a_{m2} \end{bmatrix} + \cdots + x_n \begin{bmatrix} a_{1n} \\ a_{2n} \\ \vdots \\ a_{mn} \end{bmatrix} = \begin{bmatrix} b_1 \\ b_2 \\ \vdots \\ b_m \end{bmatrix}$$

This allows all the language and theory of *vector spaces* (or more generally, *modules*) to be brought to bear.

For example, the collection of all possible linear combinations of the vectors on the left-hand side is called their *span*, and the equations have a solution just when the right-hand vector is within that span.

If every vector within that span has exactly one expression as a linear combination of the given left-hand vectors, then any solution is unique.

In any event, the span has a *basis* of linearly independent vectors that do guarantee exactly one expression; and the number of vectors in that basis (its *dimension*) cannot be larger than *m* or *n*, but it can be smaller.

This is important because if we have *m* independent vectors a solution is guaranteed regardless of the right-hand side, and otherwise not guaranteed.

Matrix Equation

The vector equation is equivalent to a matrix equation of the form $A\mathbf{x} = \mathbf{b}$ where *A* is an *m*×*n* matrix, x is a column vector with *n* entries, and b is a column vector with *m* entries.

$$A = \begin{bmatrix} a_{11} & a_{12} & \cdots & a_{1n} \\ a_{21} & a_{22} & \cdots & a_{2n} \\ \vdots & \vdots & \ddots & \vdots \\ a_{m1} & a_{m2} & \cdots & a_{mn} \end{bmatrix}, \quad \mathbf{x} = \begin{bmatrix} x_1 \\ x_2 \\ \vdots \\ x_n \end{bmatrix}, \quad \mathbf{b} = \begin{bmatrix} b_1 \\ b_2 \\ \vdots \\ b_m \end{bmatrix}$$

The number of vectors in a basis for the span is now expressed as the *rank* of the matrix.

Solution Set

A solution of a linear system is an assignment of values to the variables x_1, x_2, ..., x_n such that each of the equations is satisfied. The set of all possible solutions is called the solution set.

A linear system may behave in any one of three possible ways:

1. The system has *infinitely many solutions.*
2. The system has a single *unique solution.*
3. The system has *no solution.*

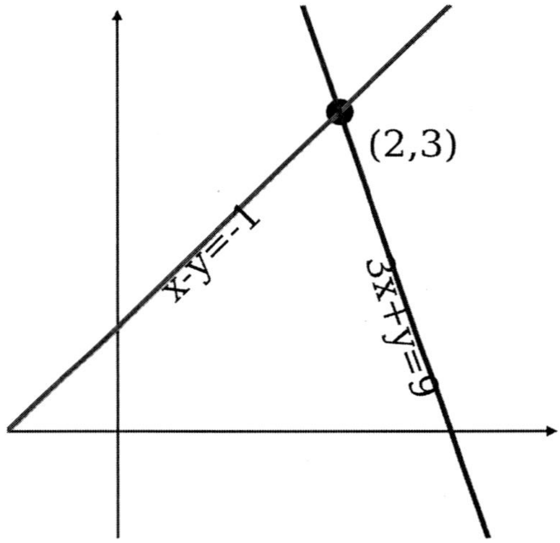

Figure: *The solution set for the equations* x – y = *–1 and* 3x + y = *9 is the single point (2, 3).*

Geometric Interpretation

For a system involving two variables (x and y), each linear equation determines a line on the xy-plane. Because a solution to a linear system must satisfy all of the equations, the solution set is the intersection of these lines, and is hence either a line, a single point, or the empty set.

For three variables, each linear equation determines a plane in three-dimensional space, and the solution set is the intersection of these planes. Thus the solution set may be a plane, a line, a single point, or the empty set.

For n variables, each linear equation determines a hyperplane in n-dimensional space. The solution set is the intersection of these hyperplanes, which may be a flat of any dimension.

General Behaviour

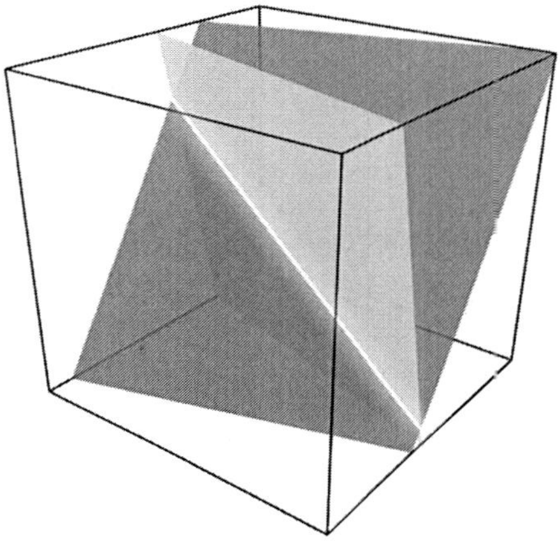

Figure: *The solution set for two equations in three variables is usually a line.*

In general, the behaviour of a linear system is determined by the relationship between the number of equations and the number of unknowns:

1. Usually, a system with fewer equations than unknowns has infinitely many solutions or sometimes unique sparse solutions (Compressed Sensing). Such a system is also known as an underdetermined system.
2. Usually, a system with the same number of equations and unknowns has a single unique solution.
3. Usually, a system with more equations than unknowns has no solution. Such a system is also known as an overdetermined system.

In the first case, the dimension of the solution set is usually equal to $n - m$, where n is the number of variables and m is the number of equations. In the first case, the dimension of the solution set is usually equal to $n - m$, where n is the number of variables and m is the number of equations.

The first system has infinitely many solutions, namely all of the points on the blue line. The second system has a single unique solution, namely the intersection of the two lines. The third system has no solutions, since the three lines share no common point. Keep in mind that the pictures above show only the most common case. It is possible for a system of two equations and two unknowns to have no solution (if the two lines are parallel), or for a system of three equations and two unknowns to be solvable (if the three lines intersect at a single point).

In general, a system of linear equations may behave differently than expected if the equations are linearly dependent, or if two or more of the equations are inconsistent.

Properties

Independence

The equations of a linear system are independent if none of the equations can be derived algebraically from the others. When the equations are independent, each equation contains new information about the variables, and removing any of the equations increases the size of the solution set.

For linear equations, logical independence is the same as linear independence.

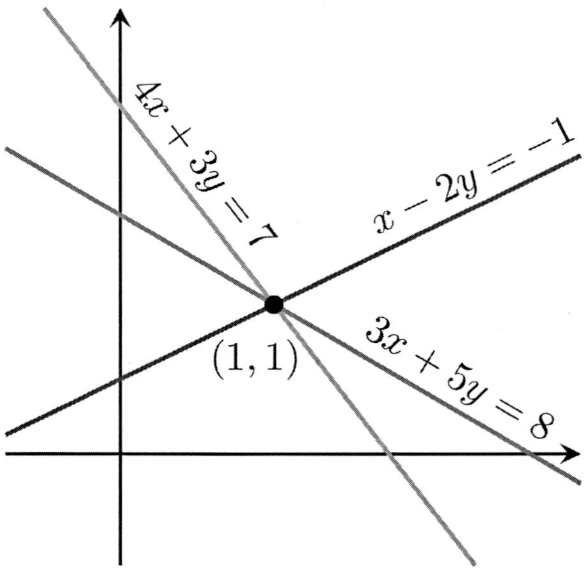

Figure: *The equations* $x - 2y = -1$*,* $3x + 5y = 8$*, and* $4x + 3y = 7$ *are not linearly independent.*

For example, the equations: $3x+2y=6$ and $6x+4y=12$ are not independent — they are the same equation when scaled by a factor of two, and they would produce identical graphs. This is an example of equivalence in a system of linear equations.

For a more complicated example, the equations

$$\begin{array}{rcrcr} x & - & 2y & = & -1 \\ 3x & + & 5y & = & 8 \\ 4x & + & 3y & = & 7 \end{array}$$

are not independent, because the third equation is the sum of the other two. Indeed, any one of these equations can be derived from the other two, and any one of the equations can be removed without affecting the solution set. The graphs of these equations are three lines that intersect at a single point.

Consistency

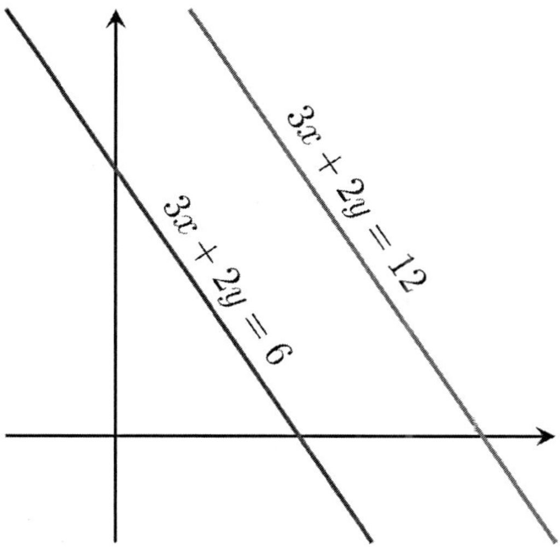

Figure: *The equations* 3x + 2y = 6 *and* 3x + 2y = 12 *are inconsistent.*

A linear system is consistent if it has a solution, and inconsistent otherwise. When the system is inconsistent, it is possible to derive a contradiction from the equations, that may always be rewritten such as the statement 0 = 1.

For example, the equations

$$3x+2y=6 \quad \text{and} \quad 3x+2y=12$$

are inconsistent. In fact, by subtracting the first equation from the second one and multiplying both sides of the result by 1/6, we get 0 = 1. The graphs of these equations on the xy-plane are a pair of parallel lines. It is possible for three linear equations to be inconsistent, even though any two of them are consistent together. For example, the equations

$$\begin{array}{ccccc} x & + & y & = & 1 \\ 2x & + & y & = & 1 \\ 3x & + & 2y & = & 3 \end{array}$$

are inconsistent. Adding the first two equations together gives $3x + 2y = 2$, which can be subtracted from the third equation to yield 0 = 1. Note that any two of these equations have a common solution. The same phenomenon can occur for any number of equations.

In general, inconsistencies occur if the left-hand sides of the equations in a system are linearly dependent, and the constant terms do not satisfy the dependence relation. A system of equations whose left-hand sides are linearly independent is always consistent.

Equivalence

Two linear systems using the same set of variables are equivalent if each of the equations in the second system can be derived algebraically from the equations in the first system, and vice-versa. Two systems are equivalent if either both are inconsistent or each equation of any of them is a linear combination of the equations of the other one. It follows that two linear systems are equivalent if and only if they have the same solution set.

Solving a Linear System

There are several algorithms for solving a system of linear equations.

Describing the Solution

When the solution set is finite, it is reduced to a single element. In this case, the unique solution is described by a sequence of equations whose left hand sides are the names of the unknowns and right hand sides are the corresponding values, for example $(x = 3, y = -2, z = 6)$. When an order on the unknowns has been fixed, for example the alphabetical order the solution may be described as a vector of values, like $(3, -2, 6)$ for the previous example.

It can be difficult to describe a set with infinite solutions. Typically, some of the variables are designated as free (or independent, or as parameters), meaning that they are allowed to take any value, while the remaining variables are dependent on the values of the free variables.

For example, consider the following system:

$$\begin{array}{ccccccc} x & + & 3y & - & 2z & = & 5 \\ 3x & + & 5y & + & 6z & = & 7 \end{array}$$

The solution set to this system can be described by the following equations:

$$x = -7z - 1 \quad \text{and} \quad y = 3z + 2.$$

Here z is the free variable, while x and y are dependent on z. Any point in the solution set can be obtained by first choosing a value for z, and then computing the corresponding values for x and y. Each free variable gives the solution space one degree of freedom, the number of which is equal to the dimension of the solution set. For example, the solution set for the above equation is a line, since a point in the solution set can be chosen by specifying the value of the parameter z. An infinite solution of higher order may describe a plane, or higher dimensional set. Different choices for the free variables may lead to different descriptions of the same solution set. For example, the solution to the above equations can alternatively be described as follows:

$$y = -\frac{3}{7}x + \frac{11}{7} \quad \text{and} \quad z = -\frac{1}{7}x - \frac{1}{7}.$$

Here x is the free variable, and y and z are dependent.

Elimination of Variables

The simplest method for solving a system of linear equations is to repeatedly eliminate variables. This method can be described as follows:

1. In the first equation, solve for one of the variables in terms of the others.
2. Plug this expression into the remaining equations. This yields a system of equations with one fewer equation and one fewer unknown.
3. Continue until you have reduced the system to a single linear equation.
4. Solve this equation, and then back-substitute until the entire solution is found.

For example, consider the following system:

$$\begin{aligned} x + 3y - 2z &= 5 \\ 3x + 5y + 6z &= 7 \\ 2x + 4y + 3z &= 8 \end{aligned}$$

Solving the first equation for x gives $x = 5 + 2z - 3y$, and plugging this into the second and third equation yields

$$\begin{aligned} -4y + 12z &= -8 \\ -2y + 7z &= -2 \end{aligned}$$

Solving the first of these equations for y yields $y = 2 + 3z$, and plugging this into the second equation yields $z = 2$. We now have:

$$\begin{aligned} x &= 5 + 2z - 3y \\ y &= 2 + 3z \\ z &= 2 \end{aligned}$$

Substituting $z = 2$ into the second equation gives $y = 8$, and substituting $z = 2$ and $y = 8$ into the first equation yields $x = -15$. Therefore, the solution set is the single point $(x, y, z) = (-15, 8, 2)$.

Cramer's Rule

Cramer's rule is an explicit formula for the solution of a system of linear equations, with each variable given by a quotient of two determinants. For example, the solution to the system

$$\begin{aligned} x + 3y - 2z &= 5 \\ 3x + 5y + 6z &= 7 \\ 2x + 4y + 3z &= 8 \end{aligned}$$

is given by

$$x = \frac{\begin{vmatrix} 5 & 3 & -2 \\ 7 & 5 & 6 \\ 8 & 4 & 3 \end{vmatrix}}{\begin{vmatrix} 1 & 3 & -2 \\ 3 & 5 & 6 \\ 2 & 4 & 3 \end{vmatrix}}, \quad y = \frac{\begin{vmatrix} 1 & 5 & -2 \\ 3 & 7 & 6 \\ 2 & 8 & 3 \end{vmatrix}}{\begin{vmatrix} 1 & 3 & -2 \\ 3 & 5 & 6 \\ 2 & 4 & 3 \end{vmatrix}}, \quad z = \frac{\begin{vmatrix} 1 & 3 & 5 \\ 3 & 5 & 7 \\ 2 & 4 & 8 \end{vmatrix}}{\begin{vmatrix} 1 & 3 & -2 \\ 3 & 5 & 6 \\ 2 & 4 & 3 \end{vmatrix}}.$$

For each variable, the denominator is the determinant of the matrix of coefficients, while the numerator is the determinant of a matrix in which one column has been replaced by the vector of constant terms.

Though Cramer's rule is important theoretically, it has little practical value for large matrices, since the computation of large determinants is somewhat cumbersome.

(Indeed, large determinants are most easily computed using row reduction.) Further, Cramer's rule has very poor numerical properties, making it unsuitable for solving even small systems reliably, unless the operations are performed in rational arithmetic with unbounded precision.

Other Methods

While systems of three or four equations can be readily solved by hand, computers are often used for larger systems. The standard algorithm for solving a system of linear equations is based on Gaussian elimination with some modifications.

Firstly, it is essential to avoid division by small numbers, which may lead to inaccurate results.

This can be done by reordering the equations if necessary, a process known as *pivoting*. Secondly, the algorithm does not exactly do Gaussian elimination, but it computes the LU decomposition of the matrix A. This is mostly an organizational tool, but it is much quicker if one has to solve several systems with the same matrix A but different vectors b.

If the matrix A has some special structure, this can be exploited to obtain faster or more accurate algorithms. For instance, systems with a symmetric positive definite matrix can be solved twice as fast with the Cholesky decomposition.

Levinson recursion is a fast method for Toeplitz matrices. Special methods exist also for matrices with many zero elements (so-called sparse matrices), which appear often in applications.

A completely different approach is often taken for very large systems, which would otherwise take too much time or memory. The idea is to start with an initial approximation to the solution (which does not have to be accurate at all), and to change this approximation in several steps to bring it closer to the true solution. Once the approximation is sufficiently accurate, this is taken to be the solution to the system. This leads to the class of iterative methods.

Homogeneous Systems

A system of linear equations is homogeneous if all of the constant terms are zero:

$$\begin{array}{ccccccc} a_{11}x_1 & + & a_{12}x_2 & +\cdots+ & a_{1n}x_n & = & 0 \\ a_{21}x_1 & + & a_{22}x_2 & +\cdots+ & a_{2n}x_n & = & 0 \\ \vdots & & \vdots & & \vdots & & \vdots \\ a_{m1}x_1 & + & a_{m2}x_2 & +\cdots+ & a_{mn}x_n & = & 0. \end{array}$$

A homogeneous system is equivalent to a matrix equation of the form

$$A\mathbf{x} = \mathbf{0}$$

where A is an $m \times n$ matrix, x is a column vector with n entries, and 0 is the zero vector with m entries.

Solution Set

Every homogeneous system has at least one solution, known as the zero solution (or trivial solution), which is obtained by assigning the value of zero to each of the variables. The solution set has the following additional properties:

1. If u and v are two vectors representing solutions to a homogeneous system, then the vector sum u + v is also a solution to the system.
2. If u is a vector representing a solution to a homogeneous system, and r is any scalar, then ru is also a solution to the system.

These are exactly the properties required for the solution set to be a linear subspace of R^n. In particular, the solution set to a homogeneous system is the same as the null space of the corresponding matrix A.

Relation to Nonhomogeneous Systems

There is a close relationship between the solutions to a linear system and the solutions to the corresponding homogeneous system:

$$A\mathbf{x} = \mathbf{b} \quad \text{and} \quad A\mathbf{x} = \mathbf{0}.$$

Specifically, if p is any specific solution to the linear system $Ax = b$, then the entire solution set can be described as

$$\{\mathbf{p} + \mathbf{v} : \mathbf{v} \text{ is any solution to } A\mathbf{x} = \mathbf{0}\}.$$

Geometrically, this says that the solution set for $Ax = b$ is a translation of the solution set for $Ax = 0$. Specifically, the flat for the first system can be obtained by translating the linear subspace for the homogeneous system by the vector p.

This reasoning only applies if the system $Ax = b$ has at least one solution. This occurs if and only if the vector b lies in the image of the linear transformation A.

Discontinuous Linear Map

In mathematics, linear maps form an important class of "simple" functions which preserve the algebraic structure of linear spaces and are often used as approximations to more general functions. If the spaces involved are also topological spaces (that is, topological vector spaces), then it makes sense to ask whether all linear maps are continuous. It turns out that for maps defined on infinite-dimensional topological vector spaces (e.g., infinite-dimensional normed spaces), the answer is generally no: there exist discontinuous linear maps. If the domain of definition is complete, such maps can be proven to exist, but the proof relies on the axiom of choice and does not provide an explicit example.

If a Linear Map is Finite Dimensional, the Linear Map is Continuous

Let X and Y be two normed spaces and f a linear map from X to Y. If X is finite-dimensional, choose a base $(e_1, e_2, \ldots, e_n)$ in X which may be taken to be unit vectors. Then,

$$f(x) = \sum_{i=1}^{n} x_i f(e_i),$$

and so by the triangle inequality,

$$\| f(x) \| = \left\| \sum_{i=1}^{n} x_i f(e_i) \right\| \le \sum_{i=1}^{n} |x_i| \| f(e_i) \| .$$

Letting

$$M = \sup_i \{\| f(e_i) \|\},$$

and using the fact that

$$\sum_{i=1}^{n} |x_i| \le C \| x \|$$

for some $C>0$ which follows from the fact that any two norms on a finite-dimensional space are equivalent, one finds

$$\| f(x) \| \le \left(\sum_{i=1}^{n} |x_i| \right) M \le CM \| x \| .$$

Thus, f is a bounded linear operator and so is continuous.

If X is infinite-dimensional, this proof will fail as there is no guarantee that the supremum M exists.

If Y is the zero space {0}, the only map between X and Y is the zero map which is trivially continuous. In all other cases, when X is infinite dimensional and Y is not the zero space, one can find a discontinuous map from X to Y.

Concrete Example

Examples of discontinuous linear maps are easy to construct in spaces that are not complete; on any Cauchy sequence of independent vectors which does not have a limit, a linear operator may grow without bound. In a sense, the linear operators are not continuous because the space has "holes".

For example, consider the space X of real-valued smooth functions on the interval [0, 1] with the uniform norm, that is,

$$\| f \| = \sup_{x \in [0,1]} | f(x) | .$$

The *derivative at a point* map, given by

$$T(f) = f'(0)$$

defined on X and with real values, is linear, but not continuous. Indeed, consider the sequence

$$f_n(x) = \frac{\sin(n^2 x)}{n}$$

for $n \geq 1$. This sequence converges uniformly to the constantly zero function, but

$$T(f_n) = \frac{n^2 \cos(n^2 \cdot 0)}{n} = n \to \infty$$

as $n \to \infty$ instead of $T(f_n) \to T(0) = 0$ which would hold for a continuous map. Note that T is real-valued, and so is actually a linear functional on X (an element of the algebraic dual space X^*). The linear map $X \to X$ which assigns to each function its derivative is similarly discontinuous. Note that although the derivative operator is not continuous, it is closed.

The fact that the domain is not complete here is important. Discontinuous operators on complete spaces require a little more work.

A Nonconstructive Example

An algebraic basis for the real numbers as a vector space over the rationals is known as a Hamel basis (note that some authors use this term in a broader sense to mean an algebraic basis of *any* vector space). Note that any two noncommensurable numbers, say 1 and π, are linearly independent. One may find a Hamel basis containing them, and define a map f from R to R so that $f(\pi) = 0$, f acts as the identity on the rest of the Hamel basis, and extend to all of R by linearity.

Let $\{r_n\}_n$ be any sequence of rationals which converges to π. Then $\lim_n f(r_n) = \pi$, but $f(\pi) = 0$. By construction, f is linear over Q (not over R), but not continuous. Note that f is also not measurable; an additive real function is linear if and only if it is measurable, so for every such function there is a Vitali set. The construction of f relies on the axiom of choice. This example can be extended into a general theorem about the existence of discontinuous linear maps on any infinite-dimensional normed space (as long as the codomain is not trivial).

General Existence Theorem

Discontinuous linear maps can be proven to exist more generally even if the space is complete. Let X and Y be normed spaces over the field K where K = R or K = C.

Assume that X is infinite-dimensional and Y is not the zero space. We will find a discontinuous linear map f from X to K, which will imply the existence of a discontinuous linear map g from X to Y given by the formula $g(x) = f(x)y_0$ where y_0 is an arbitrary nonzero vector in Y.

If X is infinite-dimensional, to show the existence of a linear functional which is not continuous then amounts to constructing f which is not bounded. For that, consider a sequence $(e_n)_n$ $(n \geq 1)$ of linearly independent vectors in X. Define

$$T(e_n) = n \| e_n \|$$

for each $n = 1, 2, \ldots$ Complete this sequence of linearly independent vectors to a vector space basis of X, and define T at the other vectors in the basis to be zero. T so defined will extend uniquely to a linear map on X, and since it is clearly not bounded, it is not continuous.

Notice that by using the fact that any set of linearly independent vectors can be completed to a basis, we implicitly used the axiom of choice, which was not needed for the concrete example in the previous section.

Axiom of Choice

As noted above, the axiom of choice (AC) is used in the general existence theorem of discontinuous linear maps. In fact, there are no constructive examples of discontinuous linear maps with complete domain (for example, Banach spaces).

In analysis as it is usually practiced by working mathematicians, the axiom of choice is always employed (it is an axiom of ZFC set theory); thus, to the analyst, all infinite dimensional topological vector spaces admit discontinuous linear maps. On the other hand, in 1970 Robert M. Solovay exhibited a model of set theory in which every set of reals is measurable. This implies that there are no discontinuous linear real functions. Clearly AC does not hold in the model.

Solovay's result shows that it is not necessary to assume that all infinite-dimensional vector spaces admit discontinuous linear maps, and there are schools of analysis which adopt a more constructivist viewpoint. For example H. G. Garnir, in searching for so-called "dream spaces" (topological vector spaces on which every linear map into a normed space is continuous), was led to adopt ZF + DC + BP (dependent choice is a weakened form and the Baire property is a negation of strong AC) as his axioms to prove the Garnir–Wright closed graph theorem which states, among other things, that any linear map from an F-space to a TVS is continuous. Going to the extreme of constructivism, there is Ceitin's theorem, which states that *every* map is continuous (where this is to be understood in an appropriate framework). Such stances are held by only a small minority of working mathematicians.

The upshot is that it is not possible to obviate the need for AC; it is consistent with set theory without AC that there are no discontinuous linear maps. A corollary is that constructible discontinuous operators such as the derivative cannot be everywhere-defined on a complete space.

Impact for Dual Spaces

The dual space of a topological vector space is the collection of continuous linear maps from the space into the underlying field. Thus the failure of some linear maps to be continuous for infinite-dimensional normed spaces implies that for these spaces, one needs to distinguish the algebraic dual space from the continuous dual space which is then a proper subset. It illustrates the fact that an extra dose of caution is needed in doing analysis on infinite-dimensional spaces as compared to finite-dimensional ones.

Beyond Normed Spaces

The argument for the existence of discontinuous linear maps on normed spaces can be generalized to all metrisable topological vector spaces, especially to all Fréchet-spaces, but there exist infinite dimensional locally convex topological vector spaces such that every functional is continuous. On the other hand, the Hahn–Banach theorem, which applies to all locally convex spaces, guarantees the existence of many continuous linear functionals, and so a large dual space. In fact, to every convex set, the Minkowski gauge associates a continuous linear functional. The upshot is that spaces with fewer convex sets have fewer functionals, and in the worst case scenario, a space may have no functionals at all other than the zero functional. This is the case for the $L^p(\mathrm{R},dx)$ spaces with $0 < p < 1$, from which it follows that these spaces are nonconvex. Note that here is indicated the Lebesgue measure on the real line. There are other L^p spaces with $0 < p < 1$ which do have nontrivial dual spaces.

Another such example is the space of real-valued measurable functions on the unit interval with quasinorm given by

$$||f|| = \int_I \frac{|f(x)|}{1+|f(x)|} dx.$$

This non-locally convex space has a trivial dual space. One can consider even more general spaces. For example, the existence of a homomorphism between complete separable metric groups can also be shown nonconstructively.

Applications

A specific application of linear maps is for geometric transformations, such as those performed in computer graphics, where the translation, rotation and scaling of 2D or 3D objects is performed by the use of a transformation matrix. Another application of these transformations is in compiler optimizations of nested-loop code, and in parallelizing compiler techniques.

Square Matrices

A *square matrix* is a matrix with the same number of rows and columns. An n-by-n matrix is known as a square matrix of order n. Any two square matrices of the same order can be added and multiplied. A square matrix A is called *invertible* or *non-singular* if there exists a matrix B such that

$$AB = I_n.$$

This is equivalent to $BA = I_n$. Moreover, if B exists, it is unique and is called the *inverse matrix* of A, denoted A^{-1}.

The entries $A_{i,i}$ form the main diagonal of a matrix. The trace, tr(A) of a square matrix A is the sum of its diagonal entries. While matrix multiplication is not commutative as mentioned above, the trace of the product of two matrices is independent of the order of the factors: tr(AB) = tr(BA).

Also, the trace of a matrix is equal to that of its transpose, i.e., $\text{tr}(A) = \text{tr}(A^T)$.

If all entries outside the main diagonal are zero, A is called a diagonal matrix. If only all entries above (below) the main diagonal are zero, A is called a lower triangular matrix (upper triangular matrix, respectively). For example, if $n = 3$, they look like

$$\begin{bmatrix} d_{11} & 0 & 0 \\ 0 & d_{22} & 0 \\ 0 & 0 & d_{33} \end{bmatrix} \text{(diagonal)},$$

$$\begin{bmatrix} l_{11} & 0 & 0 \\ l_{21} & l_{22} & 0 \\ l_{31} & l_{32} & l_{33} \end{bmatrix} \text{(lower) and } \begin{bmatrix} u_{11} & u_{12} & u_{13} \\ 0 & u_{22} & u_{23} \\ 0 & 0 & u_{33} \end{bmatrix} \text{(u}$$

Determinant

In linear algebra, the determinant is a value associated with a square matrix. It can be computed from the entries of the matrix by a specific arithmetic expression, while other ways to determine its value exist as well.

The determinant provides important information when the matrix is that of the coefficients of a system of linear equations, or when it corresponds to a linear transformation of a vector space: in the first case the system has a unique solution if and only if the determinant is nonzero, while in the second case that same condition means that the transformation has an inverse operation.

A geometric interpretation can be given to the value of the determinant of a square matrix with real entries: the absolute value of the determinant gives the scale factor by which area or volume is multiplied under the associated linear transformation, while its sign indicates whether the transformation preserves orientation. Thus a 2×2 matrix with determinant -2, when applied to a region of the plane with finite area, will transform that region into one with twice the area, while reversing its orientation.

Determinants occur throughout mathematics. The use of determinants in calculus includes the Jacobian determinant in the substitution rule for integrals of functions of several variables. They are used to define the characteristic polynomial of a matrix that is an essential tool in eigenvalue problems in linear algebra. In some cases they are used just as a compact notation for expressions that would otherwise be unwieldy to write down.

The determinant of a matrix A is denoted det(A), det A, or |A|. In the case where the matrix entries are written out in full, the determinant is denoted by surrounding the matrix entries by vertical bars instead of the brackets or parentheses of the matrix. For instance, the determinant of the matrix.

$$\begin{bmatrix} a & b & c \\ d & e & f \\ g & h & i \end{bmatrix} \text{ is written}$$

$$\begin{vmatrix} a & b & c \\ d & e & f \\ g & h & i \end{vmatrix} \text{ and has the value}$$

$$aei + bfg + cdh - ceg - bdi - afh.$$

Although most often used for matrices whose entries are real or complex numbers, the definition of the determinant only involves addition, subtraction and multiplication, and so it can be defined for square matrices with entries taken from any commutative ring.

Thus for instance the determinant of a matrix with integer coefficients will be an integer, and the matrix has an inverse with integer coefficients if and only if this determinant is 1 or −1 (these being the only invertible elements of the integers). For square matrices with entries in a non-commutative ring, for instance the quaternions, there is no unique definition for the determinant, and no definition that has all the usual properties of determinants over commutative rings.

Definition

There are various ways to define the determinant of a square matrix A, i.e. one with the same number of rows and columns. Perhaps the most natural way is expressed in terms of the columns of the matrix. If we write an n-by-n matrix in terms of its column vectors

$$A = \left[a_1, \ a_2, \ \ldots, \ a_n\right]$$

where the a_j are vectors of size n, then the determinant of A is defined so that

$$\det\left[a_1, \ \ldots, \ ba_j + cv, \ \ldots, a_n\right] =$$
$$b\det(A) + c\det\left[a_1, \ \ldots, \ v, \ \ldots, a_n\right]$$

$$\det\left[a_1, \ \ldots, \ a_j, \ a_{j+1}, \ \ldots, a_n\right]$$
$$= -\det\left[a_1, \ \ldots, \ a_{j+1}, \ a_j, \ \ldots, a_n\right]$$

$$\det(I) = 1$$

where b and c are scalars, v is any vector of size n and I is the identity matrix of size n. These properties state that the determinant is an alternating multilinear function of the columns, and they suffice to uniquely calculate the determinant of any square matrix. Provided the underlying scalars form a field (more generally, a commutative ring with unity), the definition below shows that such a function exists, and it can be shown to be unique.

Equivalently, the determinant can be expressed as a sum of products of entries of the matrix where each product has n terms and the coefficient of each product is *-1* or *1* or *0* according to a given rule: it is a polynomial expression of the matrix entries. This expression grows rapidly with the size of the matrix (an n-by-n matrix contributes $n!$ terms), so it will first be given explicitly for the case of 2-by-2 matrices and 3-by-3 matrices, followed by the rule for arbitrary size matrices, which subsumes these two cases.

Assume A is a square matrix with n rows and n columns, so that it can be written as

$$A = \begin{bmatrix} a_{1,1} & a_{1,2} & \cdots & a_{1,n} \\ a_{2,1} & a_{2,2} & \cdots & a_{2,n} \\ \vdots & \vdots & \ddots & \vdots \\ a_{n,1} & a_{n,2} & \cdots & a_{n,n} \end{bmatrix}.$$

The entries can be numbers or expressions (as happens when the determinant is used to define a characteristic polynomial); the definition of the determinant depends only on the fact that they can be added and multiplied together in a commutative manner. The determinant of A is denoted as det(A), or it can be denoted directly in terms of the matrix entries by writing enclosing bars instead of brackets:

$$\begin{vmatrix} a_{1,1} & a_{1,2} & \cdots & a_{1,n} \\ a_{2,1} & a_{2,2} & \cdots & a_{2,n} \\ \vdots & \vdots & \ddots & \vdots \\ a_{n,1} & a_{n,2} & \cdots & a_{n,n} \end{vmatrix}.$$

2-by-2 Matrices

The determinant of a 2×2 matrix is defined by

$$\begin{vmatrix} a & b \\ c & d \end{vmatrix} = ad - bc.$$

If the matrix entries are real numbers, the matrix A can be used to represent two linear mappings: one that maps the standard basis vectors to the rows of A, and one that maps them to the columns of A. In either case, the images of the basis vectors form a parallelogram that represents the image of the unit square under the mapping. The parallelogram defined by the rows of the above matrix is the one with vertices at (0,0), (a,b), (a + c, b + d), and (c,d), as shown in the accompanying diagram.

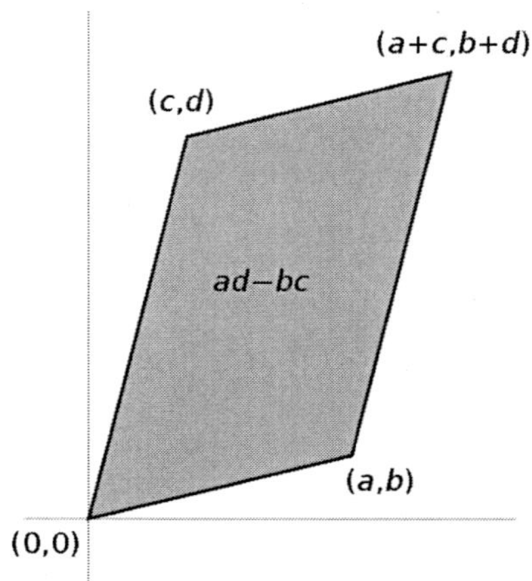

Figure: *The area of the parallelogram is the absolute value of the determinant of the matrix formed by the vectors representing the parallelogram's sides.*

The absolute value of $ad - bc$ is the area of the parallelogram, and thus represents the scale factor by which areas are transformed by A. (The parallelogram formed by the columns of A is in general a different parallelogram, but since the determinant is symmetric with respect to rows and columns, the area will be the same.)

The absolute value of the determinant together with the sign becomes the *oriented area* of the parallelogram. The oriented area is the same as the usual area, except that it is negative when the angle from the first to the second vector defining the parallelogram turns in a clockwise direction (which is opposite to the direction one would get for the identity matrix).

Thus the determinant gives the scaling factor and the orientation induced by the mapping represented by A. When the determinant is equal to one, the linear mapping defined by the matrix is equi-areal and orientation-preserving.

3-by-3 Matrices

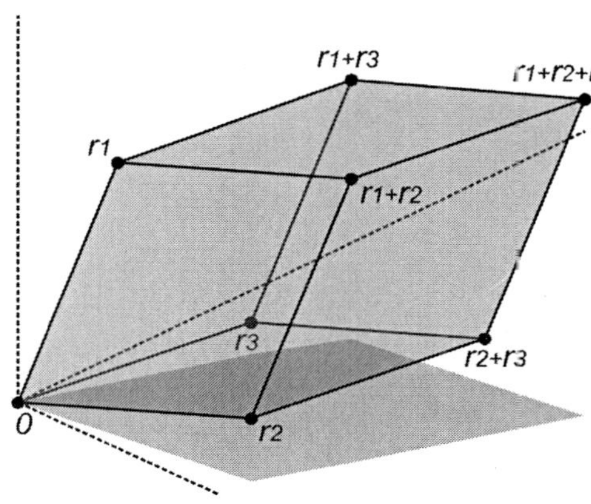

Figure: *The volume of this Parallelepiped is the absolute value of the determinant of the matrix formed by the rows r1, r2, and r3.*

The determinant of a 3×3 matrix is defined by

$$\begin{vmatrix} a & b & c \\ d & e & f \\ g & h & i \end{vmatrix} = a\begin{vmatrix} e & f \\ h & i \end{vmatrix} - b\begin{vmatrix} d & f \\ g & i \end{vmatrix} + c\begin{vmatrix} d & e \\ g & h \end{vmatrix}$$

$$= aei + bfg + cdh - ceg - bdi - afh.$$

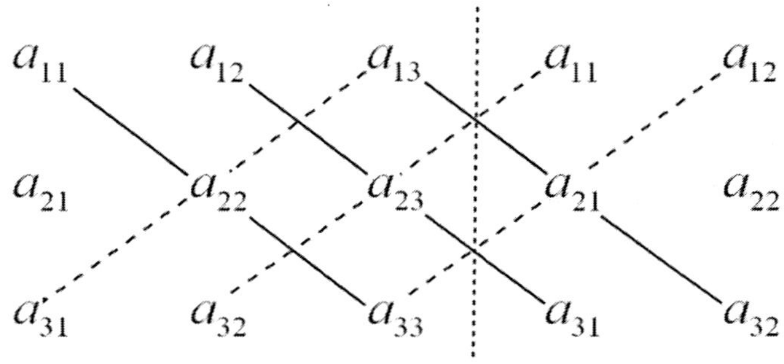

Figure: *The determinant of a 3x3 matrix can be calculated by its diagonals.*

The rule of Sarrus is a mnemonic for this formula: the sum of the products of three diagonal north-west to south-east lines of matrix elements, minus the sum of the products of three diagonal south-west to north-east lines of elements, when the copies of the first two columns of the matrix are written beside it as in the illustration.

For example, the determinant of

$$A = \begin{bmatrix} -2 & 2 & 3 \\ -1 & 1 & 3 \\ 2 & 0 & -1 \end{bmatrix}$$

is calculated using this rule:

$$\det(A) = ((-2)\cdot 1\cdot(-1)) + (2\cdot 3\cdot 2) + (3\cdot(-1)\cdot 0)$$

$$-(2\cdot 1\cdot 3) - (0\cdot 3\cdot(-2)) - ((-1)\cdot(-1)\cdot 2)$$

$$2 + 12 + 0 - 6 - 0 - 2 = 6.$$

This scheme for calculating the determinant of a 3×3 matrix does not carry over into higher dimensions. However, recently an extension to the Sarrus rule to a 4×4 matrix has been developed that requires lining up 14 columns next to each other.

Note that since 24 terms are included in the determinant of a 4×4 matrix, at least 12 columns are needed to use this method.

Properties of the Determinant

The determinant has many properties. Some basic properties of determinants are:

1. $\det(I_n) = 1$ where I_n is the $n \times n$ identity matrix.
2. $\det(A^{\mathrm{T}}) = \det(A)$.
3. $\det(A^{-1}) = \dfrac{1}{\det(A)}$.
4. $\det(AB) = \det(A)\det(B)$.
5. $\det(cA) = c^n \det(A)$ for an $n \times n$ matrix.
6. If A is a triangular matrix, i.e. $a_{i,j} = 0$ whenever $i > j$ or, alternatively, whenever $i < j$, then its determinant equals the product of the diagonal entries:

   $$\det(A) = a_{1,1}a_{2,2}\cdots a_{n,n} = \prod_{i=1}^{n} a_{i,i}.$$

   This can be deduced from some of the properties below, but it follows most easily directly from the Leibniz formula (or from the Laplace expansion), in which the identity permutation is the only one that gives a non-zero contribution.

   A number of additional properties relate to the effects on the determinant of changing particular rows or columns:

7. Viewing an $n \times n$ matrix as being composed of n columns, the determinant is an n-linear function. This means that if one column of a matrix A is written as a sum $v + w$ of two column vectors, and all other columns are left unchanged, then the determinant of A is the sum determinants of the matrices obtained from A by replacing the column by v respectively by w (and a similar relation holds when writing a column as a scalar multiple of a column vector).
8. This n-linear function is an alternating form. This means that whenever two columns of a matrix are identical, or more generally some column can be expressed as a linear combination of the other columns (i.e. the columns

of the matrix form a linearly dependent set), its determinant is 0.

Properties 1, 7 and 8 — which all follow from the Leibniz formula — completely characterize the determinant; in other words the determinant is the unique function from $n \times n$ matrices to scalars that is n-linear alternating in the columns, and takes the value 1 for the identity matrix (this characterization holds even if scalars are taken in any given commutative ring). To see this it suffices to expand the determinant by multi-linearity in the columns into a (huge) linear combination of determinants of matrices in which each column is a standard basis vector.

These determinants are either 0 (by property 8) or else ± 1 (by properties 1 and 11 below), so the linear combination gives the expression above in terms of the Levi-Civita symbol. While less technical in appearance, this characterization cannot entirely replace the Leibniz formula in defining the determinant, since without it the existence of an appropriate function is not clear. For matrices over non-commutative rings, properties 7 and 8 are incompatible for $n \geq 2$, so there is no good definition of the determinant in this setting.

Property 2 above implies that properties for columns have their counterparts in terms of rows:

9. Viewing an $n \times n$ matrix as being composed of n rows, the determinant is an n-linear function.
10. This n-linear function is an alternating form: whenever two rows of a matrix are identical, its determinant is 0.
11. Interchanging two columns of a matrix multiplies its determinant by •1. This follows from properties 7 and 8 (it is a general property of multilinear alternating maps). Iterating gives that more generally a permutation of the columns multiplies the determinant by the sign of the permutation. Similarly a permutation of the rows multiplies the determinant by the sign of the permutation.

12. Adding a scalar multiple of one column to *another* column does not change the value of the determinant. This is a consequence of properties 7 and 8: by property 7 the determinant changes by a multiple of the determinant of a matrix with two equal columns, which determinant is 0 by property 8. Similarly, adding a scalar multiple of one row to another row leaves the determinant unchanged.

These properties can be used to facilitate the computation of determinants by simplifying the matrix to the point where the determinant can be determined immediately. Specifically, for matrices with coefficients in a field, properties 11 and 12 can be used to transform any matrix into a triangular matrix, whose determinant is given by property 6; this is essentially the method of Gaussian elimination.

For example, the determinant of $A = \begin{bmatrix} -2 & 2 & -3 \\ -1 & 1 & 3 \\ 2 & 0 & -1 \end{bmatrix}$ can be computed using the following matrices:

$$B = \begin{bmatrix} -2 & 2 & -3 \\ 0 & 0 & 4.5 \\ 2 & 0 & -1 \end{bmatrix}, C = \begin{bmatrix} -2 & 2 & -3 \\ 0 & 0 & 4.5 \\ 0 & 2 & -4 \end{bmatrix}, D = \begin{bmatrix} -2 & 2 & -3 \\ 0 & 2 & -4 \\ 0 & 0 & 4.5 \end{bmatrix}.$$

Here, B is obtained from A by adding $-1/2 \times$ the first row to the second, so that $\det(A) = \det(B)$. C is obtained from B by adding the first to the third row, so that $\det(C) = \det(B)$. Finally, D is obtained from C by exchanging the second and third row, so that $\det(D) = -\det(C)$. The determinant of the (upper) triangular matrix D is the product of its entries on the main diagonal: $(-2) \cdot 2 \cdot 4.5 = -18$. Therefore $\det(A) = -\det(D) = +18$.

Multiplicativity and Matrix Groups

The determinant of a matrix product of square matrices equals the product of their determinants:

$$\det(AB) = \det(A)\det(B).$$

Thus the determinant is a *multiplicative map*. This property is a consequence of the characterization given above of the determinant as the unique n-linear alternating function of the columns with value 1 on the identity matrix, since the function $M_n(K) \to K$ that maps M ¦! det(AM) can easily be seen to be n-linear and alternating in the columns of M, and takes the value det(A) at the identity.

The formula can be generalized to (square) products of rectangular matrices, giving the Cauchy-Binet formula, which also provides an independent proof of the multiplicative property. The determinant det(A) of a matrix A is non-zero if and only if A is invertible or, yet another equivalent statement, if its rank equals the size of the matrix. If so, the determinant of the inverse matrix is given by

$$\det(A^{-1}) = \frac{1}{\det(A)}.$$

In particular, products and inverses of matrices with determinant one still have this property.

Thus, the set of such matrices (of fixed size n) form a group known as the special linear group.

More generally, the word "special" indicates the subgroup of another matrix group of matrices of determinant one. Examples include the special orthogonal group (which if n is 2 or 3 consists of all rotation matrices), and the special unitary group.

Laplace's Formula and the Adjugate Matrix

Laplace's formula expresses the determinant of a matrix in terms of its minors. The minor $M_{i,j}$ is defined to be the determinant of the (n–1)×(n–1)-matrix that results from A by removing the i-th row and the j-th column. The expression ("1)$^{i+j}M_{i,j}$ is known as cofactor. The determinant of A is given by

$$\det(A) = \sum_{j=1}^{n}(-1)^{i+j}a_{i,j}M_{i,j} = \sum_{i=1}^{n}(-1)^{i+j}a_{i,j}M_{i,j}.$$

Calculating det(A) by means of that formula is referred to as expanding the determinant along a row or column. For the example 3-by-3 matrix $A = \begin{bmatrix} -2 & 2 & -3 \\ -1 & 1 & 3 \\ 2 & 0 & -1 \end{bmatrix}$, Laplace expansion along the second column (j = 2, the sum runs over i) yields:

$$\det(A) = (-1)^{1+2}\cdot 2\cdot \det\begin{bmatrix}-1 & 3\\ 2 & -1\end{bmatrix} + (-1)^{2+2}\cdot 1\cdot \det\begin{bmatrix}-2 & -3\\ 2 & -1\end{bmatrix}$$

$$+(-1)^{3+2}\cdot 0\cdot \det\begin{bmatrix}-2 & -3\\ -1 & 3\end{bmatrix}$$

$$= (-2)\cdot((-1)\cdot(-1) - 2\cdot 3) + 1\cdot((-2)\cdot(-1) - 2\cdot(-3))$$

$$= (-2)\cdot(-5) + 8 = 18.$$

However, Laplace expansion is efficient for small matrices only.

The adjugate matrix adj(A) is the transpose of the matrix consisting of the cofactors, i.e.,

However, Laplace expansion is efficient for small matrices only.

The adjugate matrix adj(A) is the transpose of the matrix consisting of the cofactors, i.e.,

$$(\text{adj}(A))_{i,j} = (-1)^{i+j} M_{j,i}.$$

Sylvester's Determinant Theorem

Sylvester's determinant theorem states that for A, an m-by-n matrix, and B, an n-by-m matrix (so that A and B have dimensions allowing them to be multiplied in either order):

$$\det(I_m + AB) = \det(I_n + BA),$$

where I_m and I_n are the m-by-m and n-by-n identity matrices, respectively.

From this general result several consequences follow.

(a) For the case of column vector c and row vector r, each with m components, the formula allows quick calculation of the determinant of a matrix that differs from the identity matrix by a matrix of rank 1: $\det(I_m + cr) = 1 + rc$

(b) More generally, for any invertible m-by-m matrix X,

$$\det(X + AB) = \det(X)\det(I_n + BX^{-1}A)$$

(c) For a column and row vector as above, $\det(X + cr) = \det(X)(1 + rX^{-1}c)$

Properties of the Determinant in Relation to Other Notions

Eigenvalues and Eigenvectors

The eigenvectors of a square matrix are the non-zero vectors that, after being multiplied by the matrix, remain parallel to the original vector. For each eigenvector, the corresponding eigenvalue is the factor by which the eigenvector is scaled when multiplied by the matrix. The prefix eigen- is adopted from the German word "eigen" for "self" in the sense of a characteristic description. The eigenvectors are sometimes also called characteristic vectors. Similarly, the eigenvalues are also known as characteristic values.

The mathematical expression of this idea is as follows: if A is a square matrix, a non-zero vector v is an eigenvector of A if there is a scalar λ (lambda) such that

$$A\mathbf{v} = \lambda\mathbf{v}.$$

The scalar λ (lambda) is said to be the eigenvalue of A corresponding to v. An eigenspace of A is the set of all eigenvectors with the same eigenvalue together with the zero vector. However, the zero vector is not an eigenvector. These ideas are often extended to more general situations, where scalars are elements of any field, vectors are elements of any vector space, and linear transformations may or may not be represented by matrix multiplication. For example, instead of real numbers, scalars may be complex numbers; instead of arrows, vectors may be functions or frequencies; instead of matrix multiplication, linear transformations may be operators such as the derivative from calculus. These are only a few of countless examples where eigenvectors and eigenvalues are important. In such cases, the concept of *direction* loses its ordinary meaning, and is given an abstract definition. Even so, if that abstract *direction* is unchanged by a given linear transformation, the prefix "eigen" is used, as in *eigenfunction*, *eigenmode*, *eigenface*, *eigenstate*, and *eigenfrequency*. Eigenvalues and eigenvectors have many applications in both pure and applied mathematics. They are used in matrix factorization, in quantum mechanics, and in many other areas.

Definition

Prerequisites and Motivation

Eigenvectors and eigenvalues depend on the concepts of vectors and linear transformations. In the most elementary case, vectors can be thought of as arrows that have both length (or magnitude) and direction. Once a set of Cartesian coordinates is established, a vector can be described relative to that set of coordinates by a sequence of numbers. A linear transformation can be described by a square matrix. For example, in the standard coordinates of n-dimensional space, a vector can be written

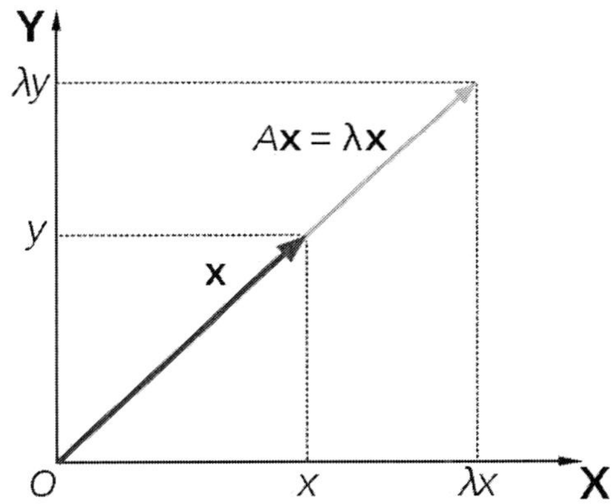

Figure: *Matrix A acts by stretching the vector x, not changing its direction, so x is an eigenvector of A.*

$$\mathbf{x} = \begin{bmatrix} x_1 \\ x_2 \\ \vdots \\ x_n \end{bmatrix}.$$

A square matrix can be written

$$A = \begin{bmatrix} a_{1,1} & a_{1,2} & \cdots & a_{1,n} \\ a_{2,1} & a_{2,2} & \cdots & a_{2,n} \\ \vdots & \vdots & \ddots & \vdots \\ a_{n,1} & a_{n,2} & \cdots & a_{n,n} \end{bmatrix}.$$

Here n is a fixed natural number.

Usually, the multiplication of a vector x by a square matrix A changes both the magnitude and the direction of the vector it acts on—but in the special case where it changes only the scale (magnitude) of the vector and leaves the direction unchanged, or switches the vector to the opposite direction, that vector is called an eigenvector of that matrix. (The term "eigenvector" is meaningless except in relation to some particular matrix.) When multiplied by a matrix, each eigenvector of that matrix changes its magnitude by a factor, called the eigenvalue corresponding to that eigenvector.

The vector x is an eigenvector of the matrix A with eigenvalue λ (lambda) if the following equation holds:

$$\mathbf{A}\mathbf{x} = \lambda\mathbf{x}.$$

This equation can be interpreted geometrically as follows: a vector x is an eigenvector if multiplication by A stretches, shrinks, leaves unchanged, flips (points in the opposite direction), flips and stretches, or flips and shrinks x. If the eigenvalue $\lambda > 1$, x is stretched by this factor. If $\lambda = 1$, the vector x is not affected at all by multiplication by A. If $0 < \lambda < 1$, x is shrunk (or compressed). The case $\lambda = 0$ means that x shrinks to a point (represented by the origin), meaning that x is in the kernel of the linear map given by A. If $\lambda < 0$ then x flips and points in the opposite direction as well as being scaled by a factor equal to the absolute value of λ.

As a special case, the identity matrix I is the matrix that leaves all vectors unchanged:

$$I\mathbf{x} = 1\mathbf{x} = \mathbf{x}.$$

Every non-zero vector x is an eigenvector of the identity matrix with eigenvalue 1.

Formal Definition

In abstract mathematics, a more general definition is given:

Let V be any vector space, let x be a non-zero vector in that vector space, and let T be a linear transformation mapping V into V. Then x is an eigenvector of T with eigenvalue λ if the following equation holds:

$$\mathbf{T}\mathbf{x} = \lambda\mathbf{x}.$$

This equation is called the *eigenvalue equation*. Note that Tx means T of x, the action of the transformation T on x, while λx means the product of the number λ times the vector x. Most, but not all authors also require x to be non-zero. The set of eigenvalues of T is sometimes called the *spectrum* of T.

Characteristic Polynomial

The eigenvalues of A are precisely the solutions λ to the equation

$$\det(A - \lambda I) = 0.$$

Here det is the determinant of the matrix formed by $A - \lambda I$ and I is the $n \times n$ identity matrix.

This equation is called the *characteristic equation* (or, less often, the secular equation) of A. For example, if A is the following matrix (a so-called diagonal matrix):

$$A = \begin{bmatrix} a_{1,1} & 0 & \cdots & 0 \\ 0 & a_{2,2} & \cdots & 0 \\ \vdots & \vdots & \ddots & 0 \\ 0 & 0 & 0 & a_{n,n} \end{bmatrix},$$

then the characteristic equation reads

$$\det(A - \lambda I) = \det\left(\begin{bmatrix} a_{1,1} & 0 & \cdots & 0 \\ 0 & a_{2,2} & \cdots & 0 \\ \vdots & \vdots & \ddots & 0 \\ 0 & 0 & 0 & a_{n,n} \end{bmatrix} - \lambda \begin{bmatrix} 1 & 0 & \cdots & 0 \\ 0 & 1 & \cdots & 0 \\ \vdots & \vdots & \ddots & 0 \\ 0 & 0 & 0 & 1 \end{bmatrix}\right)$$

$$= \det \begin{bmatrix} a_{1,1} - \lambda & 0 & \cdots & 0 \\ 0 & a_{2,2} - \lambda & \cdots & 0 \\ \vdots & \vdots & \ddots & 0 \\ 0 & 0 & 0 & a_{n,n} - \lambda \end{bmatrix}$$

$$= (a_{1,1} - \lambda)(a_{2,2} - \lambda) \cdots (a_{n,n} - \lambda) = 0$$

The solutions to this equation are the eigenvalues $\lambda_i = a_{i,i}$ ($i = 1, \ldots, n$).

Proving the aforementioned relation of eigenvalues and solutions of the characteristic equation requires some linear algebra, specifically the notion of linearly independent vectors: briefly, the eigenvalue equation for a matrix A can be expressed as

$$A\mathbf{x} - \lambda I \mathbf{x} = \mathbf{0},$$

which can be rearranged to

$$(A - \lambda I)\mathbf{x} = \mathbf{0}.$$

If there exists an inverse

$$(A - \lambda I)^{-1},$$

then both sides can be left-multiplied by it, to obtain x = 0. Therefore, if λ is such that $A - \lambda I$ is invertible, λ cannot be an eigenvalue. It can be shown that the converse holds, too: if $A - \lambda I$ is not invertible, λ is an eigenvalue. A criterion from linear algebra states that a matrix (here: $A - \lambda I$) is non-invertible if and only if its determinant is zero, thus leading to the characteristic equation.

The left-hand side of this equation can be seen (using Leibniz' rule for the determinant) to be a polynomial function in λ, whose coefficients depend on the entries of *A*. This polynomial is called the *characteristic polynomial*. Its degree is *n*, that is to say, the highest power of λ occurring in this polynomial is λ^n. At least for small matrices, the solutions of the characteristic equation (hence, the eigenvalues of *A*) can be found directly.

Moreover, it is important for theoretical purposes, such as the Cayley–Hamilton theorem. It also shows that any $n \times n$ matrix has at most *n* eigenvalues. However, the characteristic equation need not have *n* distinct solutions. In other words, there may be strictly less than *n* distinct eigenvalues. This happens for the matrix describing the shear mapping discussed below.

If the matrix has real entries, the coefficients of the characteristic polynomial are all real. However, the roots are not necessarily real; they may include complex numbers with a non-zero imaginary component.

For example, a 2×2 matrix describing a 45° rotation will not leave any non-zero vector pointing in the same direction. However, there is at least one *complex number* λ solving the characteristic equation, even if the entries of the matrix *A* are complex numbers to begin with. (This existence of such a solution is known as the fundamental theorem of algebra.) For a complex eigenvalue, the corresponding eigenvectors also have complex components.

Eigenspace

If x is an eigenvector of the matrix A with eigenvalue λ, then any scalar multiple αx is also an eigenvector of A with the same eigenvalue, since $A(\alpha x) = \alpha A x = \alpha\lambda x = \lambda(\alpha x)$. More generally, any non-zero linear combination of eigenvectors that share the same eigenvalue λ, will itself be an eigenvector with eigenvalue λ. Together with the zero vector, the eigenvectors of A with the same eigenvalue form a linear subspace called an *eigenspace* E_λ of the vector space. In case of $\dim(E_\lambda) = 1$, it is called an *eigenline* and λ is called a *scaling factor*.

Diagonalizable matrices can be decomposed into a direct sum of eigenspaces, as per the eigendecomposition of a matrix. If a matrix is not diagonalizable, then it is called defective, and, while it cannot be decomposed into eigenspaces, it can be decomposed into the more general concept of generalized eigenspaces, as discussed here.

Algebraic and Geometric Multiplicities

Given an $n \times n$ matrix A and an eigenvalue λ_i of this matrix, there are two numbers measuring, roughly speaking, the number of eigenvectors belonging to λ_i. They are called *multiplicities*: the *algebraic multiplicity* of an eigenvalue is defined as the multiplicity of the corresponding root of the characteristic polynomial. The *geometric multiplicity* of an eigenvalue is defined as the dimension of the associated eigenspace, i.e. number of linearly independent eigenvectors with that eigenvalue. Both algebraic and geometric multiplicity are integers between (including) 1 and n. The algebraic multiplicity n_i and geometric multiplicity m_i may or may not be equal, but we always have m_i dŠ n_i. The simplest case is of course when $m_i = n_i = 1$. The total number of linearly independent eigenvectors, N_x, is given by summing the geometric multiplicities

$$\sum_{i=1}^{N_\lambda} m_i = N_x.$$

Over a complex vector space, the sum of the algebraic multiplicities will equal the dimension of the vector space, but the sum of the geometric multiplicities may be smaller. In this case, it is possible that there may not be sufficient eigenvectors to span the entire space – more formally, there is no basis of eigenvectors (an eigenbasis). A matrix is diagonalizable by a suitable choice of coordinates if and only if there is an eigenbasis; if a matrix is not diagonalizable, it is said to be defective. For defective matrices, the notion of eigenvector can be generalized to generalized eigenvectors, and over an algebraically closed field a basis of *generalized* eigenvectors always exists, as follows from Jordan form.

The eigenvectors corresponding to different eigenvalues are linearly independent, meaning, in particular, that in an n-dimensional space the linear transformation A cannot have more than n eigenvalues (or eigenspaces).

All defective matrices have fewer than n distinct eigenvalues, but not all matrices with fewer than n distinct eigenvalues are defective – for example, the identity matrix is diagonalizable (and indeed diagonal in any basis), but only has the eigenvalue 1.

Given an ordered choice of linearly independent eigenvectors, especially an eigenbasis, they can be indexed by eigenvalues, *i.e.* using a double index, with $x_{i,j}$ being the j th eigenvector for the i th eigenvalue. The eigenvectors can also be indexed using the simpler notation of a single index x_k, with $k = 1, 2, ..., N_x$.

Worked Example

These concepts are explained for the matrix

$$A = \begin{bmatrix} 2 & 1 \\ 1 & 2 \end{bmatrix}.$$

The characteristic equation of this matrix reads

$$\det(A - \lambda I) = \det\begin{bmatrix} 2-\lambda & 1 \\ 1 & 2-\lambda \end{bmatrix} = 0.$$

Calculating the determinant, this yields the quadratic equation

$$\lambda^2 - 4\lambda + 3 = 0,$$

whose solutions (also called roots) are $\lambda = 1$ and $\lambda = 3$. The eigenvectors for the eigenvalue are determined by using the eigenvalue equation, which in this case reads

$$\begin{bmatrix} 2 & 1 \\ 1 & 2 \end{bmatrix} \begin{bmatrix} x \\ y \end{bmatrix} = 3 \begin{bmatrix} x \\ y \end{bmatrix}.$$

The juxtaposition at the left hand side denotes matrix multiplication. Spelling this out, this equation comparing two vectors is tantamount to a system of the following two linear equations:

$$2x + y = 3x$$
$$x + 2y = 3y.$$

Both equations reduce to the single linear equation $x = y$. That is to say, any vector of the form (x, y) with $y = x$ is an eigenvector to the eigenvalue $\lambda = 3$.

However, the vector (0, 0) is excluded. A similar calculation shows that the eigenvectors corresponding to the eigenvalue $\lambda = 1$ are given by non-zero vectors (x, y) such that $y = -x$.

For example, an eigenvector corresponding to is whereas an eigenvector corresponding to is . These vectors, placed as columns in a matrix, may be used to create a diagonalizable matrix.

Eigendecomposition

Let A be a square $n \times n$ matrix. Let $q_1 \ldots q_k$ be an eigenvector basis, i.e. an indexed set of k linearly independent eigenvectors, where k is the dimension of the space spanned by the eigenvectors of A. If $k = n$, then A can be written

$$\mathbf{A} = \mathbf{Q} \Lambda \mathbf{Q}^{-1}$$

where Q is the square $n \times n$ matrix whose i-th column is the basis eigenvector q_i of A and $\wedge$ is the diagonal matrix whose diagonal elements are the corresponding eigenvalues, i.e. $\wedge_{ii} = \lambda_i$.

Further Properties

Let A be an $n \times n$ matrix with eigenvalues $\lambda_i, i = 1, 2, \ldots, n$. Then

- Trace of A $\text{tr}(A) = \sum \lambda_i = \lambda_1 + \lambda_2 + \cdots + \lambda_n$
- Determinant of A $\det(A) = \prod \lambda_i = \lambda_1 \lambda_2 \cdots \lambda_n$
- Eigenvalues of A^k are $\lambda_1^k, \ldots, \lambda_n^k$

These first three results follow by putting the matrix in upper-triangular form, in which case the eigenvalues are on the diagonal and the trace and determinant are respectively the sum and product of the diagonal.

- If $A = A^H$, i.e., is Hermitian, every eigenvalue is real.
- Every eigenvalue of a Unitary matrix has absolute value .

Calculation

The complexity of the problem for finding roots/eigenvalues of the characteristic polynomial increases rapidly with increasing the degree of the polynomial (the dimension of the vector space). There are exact solutions for dimensions below 5, but for dimensions greater than or equal to 5 there are generally no exact solutions and one has to resort to numerical methods to find them approximately.

(In fact, since the roots of *any* polynomial can be expressed as eigenvalues of a companion matrix, the Abel–Ruffini theorem implies that there is no general algebraic solution for eigenvalues of 5×5 or larger matrices: any general eigenvalue algorithm is necessarily approximate, although in practice one can obtain any desired accuracy.)

Worse, any computational procedure that starts by computing the coefficients of the characteristic polynomial can be very inaccurate in the presence of round-off error, because the roots of a polynomial are an extremely sensitive function of the coefficients.

Efficient, accurate methods to compute eigenvalues and eigenvectors of arbitrary matrices were not known until the advent of the QR algorithm in 1961. Combining Householder transformation with LU decomposition offers better convergence than the QR algorithm. For large Hermitian sparse matrices, the Lanczos algorithm is one example of an efficient iterative method to compute eigenvalues and eigenvectors, among several other possibilities.

History

Eigenvalues are often introduced in the context of linear algebra or matrix theory. Historically, however, they arose in the study of quadratic forms and differential equations.

Euler studied the rotational motion of a rigid body and discovered the importance of the principal axes. Lagrange realized that the principal axes are the eigenvectors of the inertia matrix. In the early 19th century, Cauchy saw how their work could be used to classify the quadric surfaces, and generalized it to arbitrary dimensions. Cauchy also coined the term *racine caractéristique* (characteristic root) for what is now called *eigenvalue*; his term survives in *characteristic equation*.

Fourier used the work of Laplace and Lagrange to solve the heat equation by separation of variables in his famous 1822 book *Théorie analytique de la chaleur*. Sturm developed Fourier's ideas further and brought them to the attention of Cauchy, who combined them with his own ideas and arrived at the fact that real symmetric matrices have real eigenvalues. This was extended by Hermite in 1855 to what are now called Hermitian matrices. Around the same time, Brioschi proved that the eigenvalues of orthogonal matrices lie on the unit circle, and Clebsch found the corresponding result for skew-symmetric matrices.

Finally, Weierstrass clarified an important aspect in the stability theory started by Laplace by realizing that defective matrices can cause instability. In the meantime, Liouville studied eigenvalue problems similar to those of Sturm; the discipline that grew out of their work is now called *Sturm–Liouville theory*. Schwarz studied the first eigenvalue of Laplace's equation on general domains towards the end of the 19th century, while Poincaré studied Poisson's equation a few years later.

At the start of the 20th century, Hilbert studied the eigenvalues of integral operators by viewing the operators as infinite matrices. He was the first to use the German word *eigen* to denote eigenvalues and eigenvectors in 1904, though he may have been following a related usage by Helmholtz. For some time, the standard term in English was "proper value", but the more distinctive term "eigenvalue" is standard today.

The first numerical algorithm for computing eigenvalues and eigenvectors appeared in 1929, when Von Mises published the power method. One of the most popular methods today, the QR algorithm, was proposed independently by John G.F. Francis and Vera Kublanovskaya in 1961.

Generalizations

Left and Right Eigenvectors

The word eigenvector formally refers to the right eigenvector x_R. It is defined by the eigenvalue equation

$$Ax_R = \lambda x_R,$$

and is the most commonly used eigenvector. However, the left eigenvector x_L exists as well, and is defined by

$$x_L A = \lambda x_L.$$

The left and the right eigenvalues are the same. However, the left and right eigenvectors are usually different. But in the case of Hermitian (or real symmetric) matrix A, the left and right eigenvectors are equal.

Infinite-Dimensional Spaces and Spectral Theory

If the vector space is an infinite dimensional Banach space, the notion of eigenvalues can be generalized to the concept of spectrum. The spectrum is the set of scalars λ for which $(T - \lambda I)^{-1}$ is not defined; that is, such that $T - \lambda I$ has no bounded inverse.

Clearly if λ is an eigenvalue of T, λ is in the spectrum of T. In general, the converse is not true. There are operators on Hilbert or Banach spaces that have no eigenvectors at all. This can be seen in the following example. The bilateral shift on the Hilbert space $\ell^2(\mathbf{Z})$ (that is, the space of all sequences of scalars … a_{-1}, a_0, a_1, a_2, … such that $\cdots + |a_{-1}|^2 + |a_0|^2 + |a_1|^2 + |a_2|^2 + \cdots$ converges) has no eigenvalue but does have spectral values.

In infinite-dimensional spaces, the spectrum of a bounded operator is always nonempty. This is also true for an unbounded self adjoint operator. Via its spectral measures, the spectrum of any self adjoint operator, bounded or otherwise, can be decomposed into absolutely continuous, pure point, and singular parts.

The hydrogen atom is an example where both types of spectra appear. The eigenfunctions of the hydrogen atom Hamiltonian are called eigenstates and are grouped into two categories. The bound states of the hydrogen atom correspond to the discrete part of the spectrum (they have a discrete set of eigenvalues that can be computed by Rydberg formula) while the ionization processes are described by the continuous part (the energy of the collision/ionization is not quantized).

Eigenfunctions

A common example of such maps on infinite dimensional spaces are the action of differential operators on function spaces. As an example, on the space of infinitely differentiable functions, the process of differentiation defines a linear operator since

$$\frac{d}{dt}(af+bg)=a\frac{df}{dt}+b\frac{dg}{dt},$$

where $f(t)$ and $g(t)$ are differentiable functions, and a and b are constants.

The eigenvalue equation for linear differential operators is then a set of one or more differential equations. The eigenvectors are commonly called eigenfunctions. The simplest case is the eigenvalue equation for differentiation of a real valued function by a single real variable. We seek a function (equivalent to an infinite-dimensional vector) that, when differentiated, yields a constant times the original function. In this case, the eigenvalue equation becomes the linear differential equation

$$\frac{d}{dt}f(x)=\lambda f(x).$$

Here λ is the eigenvalue associated with the function $f(x)$. This eigenvalue equation has a solution for any value of λ. If λ is zero, the solution is

$$f(x)=A,$$

where A is any constant; if λ is non-zero, the solution is the exponential function

$$f(x)=Ae^{\lambda x}.$$

If we expand our horizons to complex valued functions, the value of λ can be any complex number. The spectrum of d/dt is therefore the whole complex plane. This is an example of a continuous spectrum.

The shape of a standing wave in a string fixed at its boundaries is an example of an eigenfunction of a differential operator. The admittable eigenvalues are governed by the length of the string and determine the frequency of oscillation. The displacement, $h(x,t)$, of a stressed rope fixed at both ends, like the vibrating strings of a string instrument, satisfies the wave equation

$$\frac{\partial^2 h}{\partial t^2} = c^2 \frac{\partial^2 h}{\partial x^2},$$

which is a linear partial differential equation, where c is the constant wave speed. The normal method of solving such an equation is separation of variables. If we assume that h can be written as the product of the form $X(x)T(t)$, we can form a pair of ordinary differential equations:

$$X'' = -\frac{\omega^2}{c^2} X \text{and} T'' = -\omega^2 T.$$

Each of these is an eigenvalue equation (the unfamiliar form of the eigenvalue is chosen merely for convenience). For any values of the eigenvalues, the eigenfunctions are given by

$$X = \sin\left(\frac{\omega x}{c} + \phi\right) \text{and} T = \sin(\omega t + \psi).$$

If we impose boundary conditions (that the ends of the string are fixed with $X(x) = 0$ at $x = 0$ and $x = L$, for example) we can constrain the eigenvalues. For those boundary conditions, we find

$\sin(\phi) = 0$, and so the phase angle $\phi = 0$ and $\sin\left(\frac{\omega L}{c}\right) = 0.$

Thus, the constant ω is constrained to take one of the values $\omega_n = \frac{nc\pi}{L}$, where n is any integer. Thus the clamped string supports a family of standing waves of the form

$$h(x,t) = \sin(n\pi x / L)\sin(\omega_n t).$$

From the point of view of our musical instrument, the frequency ω_n is the frequency of the nth harmonic, which is called the *(n-1)*st overtone.

Associative Algebras and Representation Theory

More algebraically, rather than generalizing the vector space to an infinite dimensional space, one can generalize the algebraic object that is acting on the space, replacing a single operator acting on a vector space with an algebra representation – an associative algebra acting on a module.

The study of such actions is the field of representation theory. To understand these representations, one breaks them into indecomposable representations, and, if possible, into irreducible representations; these correspond respectively to generalized eigenspaces and eigenspaces, or rather the indecomposable and irreducible components of these. While a single operator on a vector space can be understood in terms of eigenvectors – 1-dimensional invariant subspaces – in general in representation theory the building blocks (the irreducible representations) are higher-dimensional.

A closer analog of eigenvalues is given by the notion of a *weight,* with the analogs of eigenvectors and eigenspaces being *weight vectors* and *weight spaces.* For an associative algebra A over a field F, the analog of an eigenvalue is a one-dimensional representation $\lambda : A \to \mathbf{F}$ (a map of algebras; a linear functional that is also multiplicative), called the *weight,* rather than a single scalar. A map of algebras is used because if a vector is an eigenvector for two elements of an algebra, then it is also an eigenvector for any linear combination of these, and the eigenvalue is the corresponding linear combination of the eigenvalues, and likewise for multiplication.

This is related to the classical eigenvalue as follows: a single operator T corresponds to the algebra F[T] (the polynomials in T), and a map of algebras $\mathbf{F}[T] \to \mathbf{F}$ is determined by its value on the generator T; this value is the eigenvalue. A vector v on which the algebra acts by this weight (i.e., by scalar multiplication, with the scalar determined by the weight) is called a *weight vector,* and other concepts generalize similarly. The generalization of a diagonalizable matrix (having an eigenbasis) is a *weight module.*

Because a weight is a map to a field, which is commutative, the map factors through the abelianization of the algebra A – equivalently, it vanishes on the derived algebra – in terms of matrices, if v is a common eigenvector of operators T and $U,$ then $TUv = UTv$ (because in both cases it is just multiplication by scalars), so common eigenvectors of an algebra must be in the set on which the algebra acts commutatively (which is annihilated by the derived algebra).

Thus of central interest are the free commutative algebras, namely the polynomial algebras. In this particularly simple and important case of the polynomial algebra $\mathbf{F}[T_1,\ldots,T_k]$ in a set of commuting matrices, a weight vector of this algebra is a simultaneous eigenvector of the matrices, while a weight of this algebra is simply a k-tuple of scalars $\lambda = (\lambda_1,\ldots,\lambda_k)$ corresponding to the eigenvalue of each matrix, and hence geometrically to a point in k-space.

These weights – in particularly their geometry – are of central importance in understanding the representation theory of Lie algebras, specifically the finite-dimensional representations of semisimple Lie algebras.

As an application of this geometry, given an algebra that is a quotient of a polynomial algebra on k generators, it corresponds geometrically to an algebraic variety in k-dimensional space, and the weight must fall on the variety – i.e., it satisfies defining equations for the variety. This generalizes the fact that eigenvalues satisfy the characteristic polynomial of a matrix in one variable.

Applications

Schrödinger Equation

The wavefunctions associated with the bound states of an electron in a hydrogen atom can be seen as the eigenvectors of the hydrogen atom Hamiltonian as well as of the angular momentum operator. They are associated with eigenvalues interpreted as their energies (increasing downward: n=1,2,3,...) and angular momentum (increasing across: s, p, d,...). The illustration shows the square of the absolute value of the wavefunctions. Brighter areas correspond to higher probability density for a position measurement. The centre of each figure is the atomic nucleus, a proton.

An example of an eigenvalue equation where the transformation T is represented in terms of a differential operator is the time-independent Schrödinger equation in quantum mechanics:

$$H\psi_E = E\psi_E$$

where H, the Hamiltonian, is a second-order differential operator and ψ_E, the wavefunction, is one of its eigenfunctions corresponding to the eigenvalue E, interpreted as its energy.

However, in the case where one is interested only in the bound state solutions of the Schrödinger equation, one looks for within the space of square integrable functions. Since this space is a Hilbert space with a well-defined scalar product, one can introduce a basis set in which and H can be represented as a one-dimensional array and a matrix respectively. This allows one to represent the Schrödinger equation in a matrix form.

Bra-ket notation is often used in this context. A vector, which represents a state of the system, in the Hilbert space of square integrable functions is represented by . In this notation, the Schrödinger equation is:

$$H \mid \Psi_E\rangle = E \mid \Psi_E\rangle$$

where $\mid \Psi_E\rangle$ is an eigenstate of H. It is a self adjoint operator, the infinite dimensional analog of Hermitian matrices. As in the matrix case, in the equation above $H \mid \Psi_E\rangle$ is understood to be the vector obtained by application of the transformation H to $\mid \Psi_E\rangle$.

Molecular Orbitals

In quantum mechanics, and in particular in atomic and molecular physics, within the Hartree–Fock theory, the atomic and molecular orbitals can be defined by the eigenvectors of the Fock operator. The corresponding eigenvalues are interpreted as ionization potentials via Koopmans' theorem. In this case, the term eigenvector is used in a somewhat more general meaning, since the Fock operator is explicitly dependent on the orbitals and their eigenvalues. If one wants to underline this aspect one speaks of nonlinear eigenvalue problem. Such equations are usually solved by an iteration procedure, called in this case self-consistent field method. In quantum chemistry, one often represents the Hartree–Fock equation in a non-orthogonal basis set. This particular representation is a generalized eigenvalue problem called Roothaan equations.

Geology and Glaciology

In geology, especially in the study of glacial till, eigenvectors and eigenvalues are used as a method by which a mass of information of a clast fabric's constituents' orientation and dip can be summarized in a 3-D space by six numbers.

In the field, a geologist may collect such data for hundreds or thousands of clasts in a soil sample, which can only be compared graphically such as in a Tri-Plot (Sneed and Folk) diagram, or as a Stereonet on a Wulff Net. The output for the orientation tensor is in the three orthogonal (perpendicular) axes of space.

Eigenvectors output from programs such as Stereo32 are in the order $E_1 \geq E_2 \geq E_3$, with E_1 being the primary orientation of clast orientation/dip, E_2 being the secondary and E_3 being the tertiary, in terms of strength. The clast orientation is defined as the eigenvector, on a compass rose of 360°. Dip is measured as the eigenvalue, the modulus of the tensor: this is valued from 0° (no dip) to 90° (vertical). The relative values of E_1, E_2, and E_3 are dictated by the nature of the sediment's fabric. If $E_1 = E_2 = E_3$, the fabric is said to be isotropic. If $E_1 = E_2 > E_3$ the fabric is planar. If $E_1 > E_2 > E_3$ the fabric is linear.

Principal Components Analysis

The eigendecomposition of a symmetric positive semidefinite (PSD) matrix yields an orthogonal basis of eigenvectors, each of which has a nonnegative eigenvalue. The orthogonal decomposition of a PSD matrix is used in multivariate analysis, where the sample covariance matrices are PSD. This orthogonal decomposition is called principal components analysis (PCA) in statistics. PCA studies linear relations among variables. PCA is performed on the covariance matrix or the correlation matrix (in which each variable is scaled to have its sample variance equal to one). For the covariance or correlation matrix, the eigenvectors correspond to principal components and the eigenvalues to the variance explained by the principal components. Principal component analysis of the correlation matrix provides an orthonormal eigen-basis for the space of the observed data: In this basis, the largest eigenvalues correspond to the principal-components that are associated with most of the covariability among a number of observed data.

Principal component analysis is used to study large data sets, such as those encountered in data mining, chemical research, psychology, and in marketing. PCA is popular especially in psychology, in the field of psychometrics.

In Q-methodology, the eigenvalues of the correlation matrix determine the Q-methodologist's judgment of *practical* significance (which differs from the statistical significance of hypothesis testing): The factors with eigenvalues greater than 1.00 are considered practically significant, that is, as explaining an important amount of the variability in the data, while eigenvalues less than 1.00 are considered practically insignificant, as explaining only a negligible portion of the data variability. More generally, principal component analysis can be used as a method of factor analysis in structural equation modeling.

Vibration Analysis

Eigenvalue problems occur naturally in the vibration analysis of mechanical structures with many degrees of freedom. The eigenvalues are used to determine the natural frequencies (or eigenfrequencies) of vibration, and the eigenvectors determine the shapes of these vibrational modes. In particular, undamped vibration is governed by

$$m\ddot{x} + kx = 0$$

or

$$m\ddot{x} = -kx$$

that is, acceleration is proportional to position (i.e., we expect x to be sinusoidal in time). In n dimensions, m becomes a mass matrix and k a stiffness matrix. Admissible solutions are then a linear combination of solutions to the generalized eigenvalue problem

$$-kx = \omega^2 mx$$

where ω^2 is the eigenvalue and ω is the angular frequency. Note that the principal vibration modes are different from the principal compliance modes, which are the eigenvectors of k alone. Furthermore, damped vibration, governed by

$$m\ddot{x} + c\dot{x} + kx = 0$$

leads to what is called a so-called quadratic eigenvalue problem,

$$(\omega^2 m + \omega c + k)x = 0$$

This can be reduced to a generalized eigenvalue problem by clever algebra at the cost of solving a larger system.

The orthogonality properties of the eigenvectors allows decoupling of the differential equations so that the system can be represented as linear summation of the eigenvectors. The eigenvalue problem of complex structures is often solved using finite element analysis, but neatly generalize the solution to scalar-valued vibration problems.

Eigenfaces

In image processing, processed images of faces can be seen as vectors whose components are the brightnesses of each pixel. The dimension of this vector space is the number of pixels. The eigenvectors of the covariance matrix associated with a large set of normalized pictures of faces are called eigenfaces; this is an example of principal components analysis. They are very useful for expressing any face image as a linear combination of some of them. In the facial recognition branch of biometrics, eigenfaces provide a means of applying data compression to faces for identification purposes. Research related to eigen vision systems determining hand gestures has also been made.

Similar to this concept, eigenvoices represent the general direction of variability in human pronunciations of a particular utterance, such as a word in a language. Based on a linear combination of such eigenvoices, a new voice pronunciation of the word can be constructed. These concepts have been found useful in automatic speech recognition systems, for speaker adaptation.

Tensor of Inertia

In mechanics, the eigenvectors of the inertia tensor define the principal axes of a rigid body. The tensor of inertia is a key quantity required to determine the rotation of a rigid body around its centre of mass.

Stress Tensor

In solid mechanics, the stress tensor is symmetric and so can be decomposed into a diagonal tensor with the eigenvalues on the diagonal and eigenvectors as a basis. Because it is diagonal, in this orientation, the stress tensor has no shear components; the components it does have are the principal components.

Eigenvalues of a Graph

In spectral graph theory, an eigenvalue of a graph is defined as an eigenvalue of the graph's adjacency matrix A, or (increasingly) of the graph's Laplacian matrix, which is either $T-A$ (sometimes called the Combinatorial Laplacian) or $I-T^{-1/2}AT^{-1/2}$ (sometimes called the Normalized Laplacian), where T is a diagonal matrix with $T_{v,v}$ equal to the degree of vertex v, and in $T^{-1/2}$, the vth diagonal entry is $\deg(v)^{-1/2}$. The kth principal eigenvector of a graph is defined as either the eigenvector corresponding to the kth largest or k^{th} smallest eigenvalue of the Laplacian. The first principal eigenvector of the graph is also referred to merely as the principal eigenvector.

The principal eigenvector is used to measure the centrality of its vertices. An example is Google's PageRank algorithm. The principal eigenvector of a modified adjacency matrix of the World Wide Web graph gives the page ranks as its components.

This vector corresponds to the stationary distribution of the Markov chain represented by the row-normalized adjacency matrix; however, the adjacency matrix must first be modified to ensure a stationary distribution exists. The second smallest eigenvector can be used to partition the graph into clusters, via spectral clustering. Other methods are also available for clustering.

Cramer's Rule

For a matrix equation

$$Ax = b$$

the solution is given by Cramer's rule:

$$x_i = \frac{\det(A_i)}{\det(A)} \qquad i = 1, \ldots, n$$

where A_i is the matrix formed by replacing the i-th column of A by the column vector b. This follows immediately by column expansion of the determinant, i.e.

$$\begin{aligned}\det(A_i) &= \det\left[a_1, \quad \ldots, \quad b, \quad \ldots, \quad a_n\right] \\ &= \sum_{j=1}^{n} x_j \det\left[a_1, \quad \ldots, a_{i-1}, \quad a_j, \quad a_{i+1}, \quad \ldots, \quad a_n\right] \\ &= x_i \det(A)\end{aligned}$$

where the vectors a_j are the columns of A. The rule is also implied by the identity

$$A\,\mathrm{adj}(A) = \mathrm{adj}(A)A = \det(A)I_n.$$

It has recently been shown that Cramer's rule can be implemented in $O(n^3)$ time, which is comparable to more common methods of solving systems of linear equations, such as LU, QR, or singular value decomposition.

Block Matrices

Suppose A, B, C, and D are $n\times n$-, $n\times m$-, $m\times n$-, and $m\times m$-matrices, respectively. Then

$$\det\begin{pmatrix} A & 0 \\ C & D \end{pmatrix} = \det\begin{pmatrix} A & B \\ 0 & D \end{pmatrix} = \det(A)\det(D).$$

This can be seen from the Leibniz formula or by induction on n. When A is invertible, employing the following identity

$$\begin{pmatrix} A & B \\ C & D \end{pmatrix} = \begin{pmatrix} A & 0 \\ C & I \end{pmatrix}\begin{pmatrix} I & A^{-1}B \\ 0 & D-CA^{-1}B \end{pmatrix}$$

leads to

$$\det\begin{pmatrix} A & B \\ C & D \end{pmatrix} = \det(A)\det(D-CA^{-1}B).$$

When D is invertible, a similar identity with $\det(D)$ factored out can be derived analogously, that is,

$$\det\begin{pmatrix} A & B \\ C & D \end{pmatrix} = \det(D)\det(A-BD^{-1}C).$$

When the blocks are square matrices of the same order further formulas hold. For example, if C and D commute (i.e., $CD = DC$), then the following formula comparable to the determinant of a 2-by-2 matrix holds:

$$\det\begin{pmatrix} A & B \\ C & D \end{pmatrix} = \det(AD - BC).$$

Derivative

By definition, e.g., using the Leibniz formula, the determinant of real (or analogously for complex) square matrices is a polynomial function from $R^{n \times n}$ to R. As such it is everywhere differentiable. Its derivative can be expressed using Jacobi's formula:

$$\frac{\mathrm{d}\det(A)}{\mathrm{d}\alpha} = \operatorname{tr}\left(\operatorname{adj}(A)\frac{\mathrm{d}A}{\mathrm{d}\alpha}\right).$$

where adj(A) denotes the adjugate of A. In particular, if A is invertible, we have

$$\frac{\mathrm{d}\det(A)}{\mathrm{d}\alpha} = \det(A)\operatorname{tr}\left(A^{-1}\frac{\mathrm{d}A}{\mathrm{d}\alpha}\right).$$

Expressed in terms of the entries of *A*, these are

$$\frac{\partial \det(A)}{\partial A_{ij}} = \operatorname{adj}(A)_{ji} = \det(A)(A^{-1})_{ji}.$$

Yet another equivalent formulation is

$$\det(A + \epsilon X) - \det(A) = \operatorname{tr}(\operatorname{adj}(A)X)\epsilon + O(\epsilon^2)$$
$$= \det(A)\operatorname{tr}(A^{-1}X)\epsilon + O(\epsilon^2)$$

using big O notation. The special case where $A = I$, the identity matrix, yields

$$\det(I + \epsilon X) = 1 + \operatorname{tr}(X)\epsilon + O(\epsilon^2).$$

This identity is used in describing the tangent space of certain matrix Lie groups.

If the matrix A is written as $A = \begin{bmatrix}\mathbf{a} & \mathbf{b} & \mathbf{c}\end{bmatrix}$ where a, b, c are vectors, then the gradient over one of the three vectors may be written as the cross product of the other two:

$$\begin{aligned} \nabla_{\mathbf{a}} \det(A) &= \mathbf{b} \times \mathbf{c} \\ \nabla_{\mathbf{b}} \det(A) &= \mathbf{c} \times \mathbf{a} \\ \nabla_{\mathbf{c}} \det(A) &= \mathbf{a} \times \mathbf{b}. \end{aligned}$$

Chapter 5

Basic Examples of Matrices

The set $M_R(n,n)$ of $n \times n$ matrices over a commutative ring R is itself a ring under matrix addition and multiplication. The group of units of $M_R(n,n)$ is called the general linear group of $n \times n$ matrices over the ring R and is denoted $GL_n(R)$ or $GL(n,R)$. All matrix groups are subgroups of some general linear group.

Classical Group

In mathematics, the classical Lie groups are four infinite families of Lie groups closely related to the symmetries of Euclidean spaces. Their finite analogues are the classical groups of Lie type. The term was coined by Hermann Weyl. Contrasting with the classical Lie groups are the exceptional Lie groups, which share their abstract properties, but not their familiarity. Sometimes classical groups are discussed in the restricted setting of compact groups, a formulation which makes their representation theory and algebraic topology easiest to handle. It does however exclude the general linear group.

Relationship with Bilinear Forms

The unifying feature of classical Lie groups is that they are close to the isometry groups of certain bilinear or sesquilinear forms. The four series are labelled by the Dynkin diagram attached to them, with subscript $n \geq 1$. The families may be represented as follows:

- A_n = SU(n + 1), the special unitary group of unitary n+1-by-n+1 complex matrices with determinant 1.
- B_n = SO($2n$ + 1), the special orthogonal group of orthogonal ($2n$ + 1)-by-($2n$ + 1) real matrices with determinant 1.

- C_n = Sp(n), the symplectic group of n-by-n quaternionic matrices that preserve the usual inner product on H^n.
- D_n = SO($2n$), the special orthogonal group of orthogonal $2n$-by-$2n$ real matrices with determinant 1.

For certain purposes it is also natural to drop the condition that the determinant be 1 and consider unitary groups and (disconnected) orthogonal groups.

The table lists the so-called connected compact real forms of the groups; they have closely related complex analogues and various non-compact forms, for example, together with compact orthogonal groups one considers indefinite orthogonal groups.

The Lie algebras corresponding to these groups are known as the *classical Lie algebras*. Viewing a classical group G as a subgroup of GL(n) via its definition as automorphisms of a vector space preserving some involution provides a representation of G called the standard representation.

Classical Groups Over General Fields or Rings

Classical groups, more broadly considered in algebra, provide particularly interesting matrix groups. When the ring of coefficients of the matrix group is the real number or complex number field, these groups are just certain of the classical Lie groups.

When the underlying ring is a finite field the classical groups are groups of Lie type. These groups play an important role in the classification of finite simple groups. Considering their abstract group theory, many linear groups have a "special" subgroup, usually consisting of the elements of determinant 1 (for orthogonal groups in characteristic 2 it consists of the elements of Dickson invariant 0), and most of them have associated "projective" quotients, which are the quotients by the centre of the group.

The word "general" in front of a group name usually means that the group is allowed to multiply some sort of form by a constant, rather than leaving it fixed.

The subscript n usually indicates the dimension of the module on which the group is acting. Caveat: this notation clashes somewhat with the n of Dynkin diagrams, which is the rank.

General and Special Linear Groups

The general linear group $GL_n(R)$ is the group of all R-linear automorphisms of R^n. There is a subgroup: the special linear group $SL_n(R)$, and their quotients: the projective general linear group $PGL_n(R) = GL_n(R)/Z(GL_n(R))$ and the projective special linear group $PSL_n(R) = SL_n(R)/Z(SL_n(R))$. The projective special linear group $PSL_n(R)$ over a field R is simple for $n \geq 2$, except for the two cases when $n = 2$ and the field has order 2 or 3.

Unitary Groups

The unitary group $U_n(R)$ is a group preserving a sesquilinear form on a module. There is a subgroup, the special unitary group $SU_n(R)$ and their quotients the projective unitary group $PU_n(R) = U_n(R)/Z(U_n(R))$ and the projective special unitary group $PSU_n(R) = SU_n(R)/Z(SU_n(R))$

Symplectic Groups

The symplectic group $Sp_{2n}(R)$ preserves a skew symmetric form on a module. It has a quotient, the projective symplectic group $PSp_{2n}(R)$. The general symplectic group $GSp_{2n}(R)$ consists of the automorphisms of a module multiplying a skew symmetric form by some invertible scalar. The projective symplectic group $PSp_{2n}(R)$ over a finite field R is simple for $n \geq 1$, except for the two cases when $n = 1$ and the field has order 2 or 3.

Orthogonal Groups

The orthogonal group $O_n(R)$ preserves a non-degenerate quadratic form on a module. There is a subgroup, the special orthogonal group $SO_n(R)$ and quotients, the projective orthogonal group $PO_n(R)$, and the projective special orthogonal group $PSO_n(R)$. (In characteristic 2 the determinant is always 1, so the special orthogonal group is often defined as the subgroup of elements of Dickson invariant 1.)

There is a nameless group often denoted by $\Omega_n(R)$ consisting of the elements of the orthogonal group of elements of spinor norm 1, with corresponding subgroup and quotient groups $S\Omega_n(R)$, $P\Omega_n(R)$, $PS\Omega_n(R)$.

(For positive definite quadratic forms over the reals, the group Ω happens to be the same as the orthogonal group, but in general it is smaller.)

There is also a double cover of $\Omega_n(R)$, called the pin group $Pin_n(R)$, and it has a subgroup called the spin group $Spin_n(R)$. The general orthogonal group $GO_n(R)$ consists of the automorphisms of a module multiplying a quadratic form by some invertible scalar.

Notational Conventions

Finite Groups as Matrix Groups

Every finite group is isomorphic to some matrix group. This is similar to Cayley's theorem which states that every finite group is isomorphic to some permutation group. Since the isomorphism property is transitive one need only consider how to form a matrix group from a permutation group.

Let G be a permutation group on n points ($\Omega = \{1,2,\ldots,n\}$) and let $\{g_1,\ldots,g_k\}$ be a generating set for G. The general linear group $GL_n(C)$ of $n \times n$ matrices over the complex numbers acts naturally on the vector space C^n. Let $B=\{b_1,\ldots,b_n\}$ be the standard basis for C^n. For each g_i let M_i in $GL_n(C)$ be the matrix which sends each b_j to $b_{gi(j)}$. That is, if the permutation g_i sends the point j to k then M_i sends the basis vector b_j to b_k.

Let M be the subgroup of $GL_n(C)$ generated by $\{M_1,\ldots,M_k\}$. The action of G on Ω is then precisely the same as the action of M on B. It can be proved that the function taking each g_i to M_i extends to an isomorphism and thus every group is isomorphic to a matrix group.

Note that the field (C in the above case) is irrelevant since M contains only elements with entries 0 or 1. One can just as easily perform the construction for an arbitrary field since the elements 0 and 1 exist in every field.

As an example, let $G = S_3$, the symmetric group on 3 points. Let $g_1 = (1,2,3)$ and $g_2 = (1,2)$. Then

$$M_1 = \begin{bmatrix} 0 & 1 & 0 \\ 0 & 0 & 1 \\ 1 & 0 & 0 \end{bmatrix}$$

$$M_2 = \begin{bmatrix} 0 & 1 & 0 \\ 1 & 0 & 0 \\ 0 & 0 & 1 \end{bmatrix}$$

Notice that $M_1b_1 = b_2$, $M_1b_2 = b_3$ and $M_1b_3 = b_1$. Likewise, $M_2b_1 = b_2$, $M_2b_2 = b_1$ and $M_2b_3 = b_3$.

Representation Theory and Character Theory

Linear transformations and matrices are (generally speaking) well-understood objects in mathematics and have been used extensively in the study of groups. In particular representation theory studies homomorphisms from a group into a matrix group and character theory studies homomorphisms from a group into a field given by the trace of a representation.

Examples

- In 2000 a longstanding conjecture was resolved when it was shown that the braid groups B_n are linear for all n.
- Infinite matrices
- It is also possible to consider matrices with infinitely many rows and/or columns even if, being infinite objects, one cannot write down such matrices explicitly. All that matters is that for every element in the set indexing rows, and every element in the set indexing columns, there is a well-defined entry (these index sets need not even be subsets of the natural numbers). The basic operations of addition, subtraction, scalar multiplication and transposition can still be defined without problem; however matrix multiplication may involve infinite summations to define the resulting entries, and these are not defined in general.
- If R is any ring with unity, then the ring of endomorphisms of $M = \bigoplus_{i \in I} R$ as a right R module is isomorphic to the ring of column finite matrices $\mathbb{CFM}_I(R)$ whose entries are indexed by $I \times I$, and whose columns each contain only finitely many nonzero entries. The endomorphisms of M considered as a left R module result in an analogous object, the row finite matrices $\mathbb{RFM}_I(R)$ whose rows each only have finitely many nonzero entries.

- If infinite matrices are used to describe linear maps, then only those matrices can be used all of whose columns have but a finite number of nonzero entries, for the following reason. For a matrix A to describe a linear map f: $V \to W$, bases for both spaces must have been chosen; recall that by definition this means that every vector in the space can be written uniquely as a (finite) linear combination of basis vectors, so that written as a (column) vector v of coefficients, only finitely many entries v_i are nonzero. Now the columns of A describe the images by f of individual basis vectors of V in the basis of W, which is only meaningful if these columns have only finitely many nonzero entries. There is no restriction on the rows of A however: in the product $A \cdot v$ there are only finitely many nonzero coefficients of v involved, so every one of its entries, even if it is given as an infinite sum of products, involves only finitely many nonzero terms and is therefore well defined. Moreover this amounts to forming a linear combination of the columns of A that effectively involves only finitely many of them, whence the result has only finitely many nonzero entries, because each of those columns do. One also sees that products of two matrices of the given type is well defined (provided as usual that the column-index and row-index sets match), is again of the same type, and corresponds to the composition of linear maps.
- If R is a normed ring, then the condition of row or column finiteness can be relaxed. With the norm in place, absolutely convergent series can be used instead of finite sums. For example, the matrices whose column sums are absolutely convergent sequences form a ring. Analogously of course, the matrices whose row sums are absolutely convergent series also form a ring.
- In that vein, infinite matrices can also be used to describe operators on Hilbert spaces, where convergence and continuity questions arise, which again results in certain constraints that have to be imposed. However, the explicit point of view of matrices tends to obfuscate the matter, and the abstract and more powerful tools of functional analysis can be used instead.

- Empty matrices
- An *empty matrix* is a matrix in which the number of rows or columns (or both) is zero. Empty matrices help dealing with maps involving the zero vector space. For example, if A is a 3-by-0 matrix and B is a 0-by-3 matrix, then AB is the 3-by-3 zero matrix corresponding to the null map from a 3-dimensional space V to itself, while BA is a 0-by-0 matrix. There is no common notation for empty matrices, but most computer algebra systems allow creating and computing with them. The determinant of the 0-by-0 matrix is 1 as follows from regarding the empty product occurring in the Leibniz formula for the determinant as 1. This value is also consistent with the fact that the identity map from any finite dimensional space to itself has determinant 1, a fact that is often used as a part of the characterization of determinants.

Applications

There are numerous applications of matrices, both in mathematics and other sciences. Some of them merely take advantage of the compact representation of a set of numbers in a matrix. For example, in game theory and economics, the payoff matrix encodes the payoff for two players, depending on which out of a given (finite) set of alternatives the players choose. Text mining and automated thesaurus compilation makes use of document-term matrices such as tf-idf to track frequencies of certain words in several documents.

Complex numbers can be represented by particular real 2-by-2 matrices via

$$a + ib \leftrightarrow \begin{bmatrix} a & -b \\ b & a \end{bmatrix},$$

under which addition and multiplication of complex numbers and matrices correspond to each other. For example, 2-by-2 rotation matrices represent the multiplication with some complex number of absolute value 1, as above. A similar interpretation is possible for quaternions.

Early encryption techniques such as the Hill cipher also used matrices. However, due to the linear nature of matrices, these codes are comparatively easy to break. Computer graphics uses matrices both to represent objects and to calculate transformations of objects using affine rotation matrices to accomplish tasks such as projecting a three-dimensional object onto a two-dimensional screen, corresponding to a theoretical camera observation. Matrices over a polynomial ring are important in the study of control theory.

Chemistry makes use of matrices in various ways, particularly since the use of quantum theory to discuss molecular bonding and spectroscopy. Examples are the overlap matrix and the Fock matrix used in solving the Roothaan equations to obtain the molecular orbitals of the Hartree–Fock method.

Symmetries and Rransformations in Physics

Linear transformations and the associated symmetries play a key role in modern physics. For example, elementary particles in quantum field theory are classified as representations of the Lorentz group of special relativity and, more specifically, by their behaviour under the spin group. Concrete representations involving the Pauli matrices and more general gamma matrices are an integral part of the physical description of fermions, which behave as spinors.

For the three lightest quarks, there is a group-theoretical representation involving the special unitary group SU(3); for their calculations, physicists use a convenient matrix representation known as the Gell-Mann matrices, which are also used for the SU(3) gauge group that forms the basis of the modern description of strong nuclear interactions, quantum chromodynamics. The Cabibbo–Kobayashi–Maskawa matrix, in turn, expresses the fact that the basic quark states that are important for weak interactions are not the same as, but linearly related to the basic quark states that define particles with specific and distinct masses.

Linear Combinations of Quantum States

The first model of quantum mechanics (Heisenberg, 1925) represented the theory's operators by infinite-dimensional matrices acting on quantum states. This is also referred to as matrix mechanics. One particular example is the density matrix that characterizes the "mixed" state of a quantum system as a linear combination of elementary, "pure" eigenstates.

Another matrix serves as a key tool for describing the scattering experiments that form the cornerstone of experimental particle physics: Collision reactions such as occur in particle accelerators, where non-interacting particles head towards each other and collide in a small interaction zone, with a new set of non-interacting particles as the result, can be described as the scalar product of outgoing particle states and a linear combination of ingoing particle states. The linear combination is given by a matrix known as the S-matrix, which encodes all information about the possible interactions between particles.

Normal Modes

A general application of matrices in physics is to the description of linearly coupled harmonic systems. The equations of motion of such systems can be described in matrix form, with a mass matrix multiplying a generalized velocity to give the kinetic term, and a force matrix multiplying a displacement vector to characterize the interactions. The best way to obtain solutions is to determine the system's eigenvectors, its normal modes, by diagonalizing the matrix equation. Techniques like this are crucial when it comes to the internal dynamics of molecules: the internal vibrations of systems consisting of mutually bound component atoms. They are also needed for describing mechanical vibrations, and oscillations in electrical circuits.

Geometrical Optics

Geometrical optics provides further matrix applications. In this approximative theory, the wave nature of light is neglected. The result is a model in which light rays are indeed geometrical rays. If the deflection of light rays by optical elements is small, the action of a lens or reflective element on a given light ray can be expressed as multiplication of a two-component vector with a two-by-two matrix called ray transfer matrix: the vector's components are the light ray's slope and its distance from the optical axis, while the matrix encodes the properties of the optical element. Actually, there are two kinds of matrices, viz. a *refraction matrix* describing the refraction at a lens surface, and a *translation matrix*, describing the translation of the plane of reference to the next refracting surface, where another refraction matrix applies. The optical system, consisting of a combination of lenses and/or reflective elements, is simply described by the matrix resulting from the product of the components' matrices.

Electronics

Traditional mesh analysis in electronics leads to a system of linear equations that can be described with a matrix.

The behaviour of many electronic components can be described using matrices. Let A be a 2-dimensional vector with the component's input voltage v_1 and input current i_1 as its elements, and let B be a 2-dimensional vector with the component's output voltage v_2 and output current i_2 as its elements. Then the behaviour of the electronic component can be described by $B = H \cdot A$, where H is a 2×2 matrix containing one impedance element (h_{12}), one admittance element (h_{21}) and two dimensionless elements (h_{11} and h_{22}). Calculating a circuit now reduces to multiplying matrices.

History

Matrices have a long history of application in solving linear equations. The Chinese text *The Nine Chapters on the Mathematical Art* (*Jiu Zhang Suan Shu*), from between 300 BC and AD 200, is the first example of the use of matrix methods to solve simultaneous equations, including the concept of determinants, over 1000 years before its publication by the Japanese mathematician Seki in 1683 and the German mathematician Leibniz in 1693. Cramer presented his rule in 1750.

Early matrix theory emphasized determinants more strongly than matrices and an independent matrix concept akin to the modern notion emerged only in 1858, with Cayley's *Memoir on the theory of matrices.* The term "matrix" (Latin for "womb", derived from *mater*—mother) was coined by Sylvester, who understood a matrix as an object giving rise to a number of determinants today called minors, that is to say, determinants of smaller matrices that derive from the original one by removing columns and rows. In a 1851 paper, Sylvester explains:

I have in previous papers defined a "Matrix" as a rectangular array of terms, out of which different systems of determinants may be engendered as from the womb of a common parent.

The study of determinants sprang from several sources. Number-theoretical problems led Gauss to relate coefficients of quadratic forms, i.e., expressions such as $x^2 + xy - 2y^2$, and linear maps in three dimensions to matrices. Eisenstein further developed these notions, including the remark that, in modern parlance, matrix products are non-commutative. Cauchy was the first to prove general statements about determinants, using as definition of the determinant of a matrix $A = [a_{i,j}]$ the following: replace the powers a_j^k by a_{jk} in the polynomial

$$a_1 a_2 \cdots a_n \prod_{i<j} (a_j - a_i) ,$$

where Π denotes the product of the indicated terms. He also showed, in 1829, that the eigenvalues of symmetric matrices are real. Jacobi studied "functional determinants"—later called Jacobi determinants by Sylvester—which can be used to describe geometric transformations at a local (or infinitesimal) level, Kronecker's *Vorlesungen über die Theorie der Determinanten* and Weierstrass' *Zur Determinantentheorie*, both published in 1903, first treated determinants axiomatically, as opposed to previous more concrete approaches such as the mentioned formula of Cauchy. At that point, determinants were firmly established.

Many theorems were first established for small matrices only, for example the Cayley–Hamilton theorem was proved for 2×2 matrices by Cayley in the aforementioned memoir, and by Hamilton for 4×4 matrices. Frobenius, working on bilinear forms, generalized the theorem to all dimensions (1898). Also at the end of the 19th century the Gauss–Jordan elimination (generalizing a special case now known as Gauss elimination) was established by Jordan. In the early 20th century, matrices attained a central role in linear algebra partially due to their use in classification of the hypercomplex number systems of the previous century.

The inception of matrix mechanics by Heisenberg, Born and Jordan led to studying matrices with infinitely many rows and columns. Later, von Neumann carried out the mathematical formulation of quantum mechanics, by further developing functional analytic notions such as linear operators on Hilbert spaces, which, very roughly speaking, correspond to Euclidean space, but with an infinity of independent directions.

Other Historical Usages of the Word "Matrix" in Mathematics

The word has been used in unusual ways by at least two authors of historical importance.

Bertrand Russell and Alfred North Whitehead in their *Principia Mathematica* (1910–1913) use the word matrix in the context of their Axiom of reducibility. They proposed this axiom as a means to reduce any function to one of lower type, successively, so that at the "bottom" (0 order) the function is identical to its extension:

"Let us give the name of *matrix* to any function, of however many variables, which does not involve any apparent variables. Then any possible function other than a matrix is derived from a matrix by means of generalization, i.e., by considering the proposition which asserts that the function in question is true with all possible values or with some value of one of the arguments, the other argument or arguments remaining undetermined".

For example a function $\Phi(x, y)$ of two variables x and y can be reduced to a *collection* of functions of a single variable, e.g., y, by "considering" the function for all possible values of "individuals" a_i substituted in place of variable x. And then the resulting collection of functions of the single variable y, i.e., $\forall a_i : \Phi(a_i, y)$, can be reduced to a "matrix" of values by "considering" the function for all possible values of "individuals" b_i substituted in place of variable y:

$$\forall b_j \forall a_i : \phi(a_i, b_j)$$

Alfred Tarski in his 1946 *Introduction to Logic* used the word "matrix" synonymously with the notion of truth table as used in mathematical logic.

Random Matrix

In probability theory and mathematical physics, a random matrix is a matrix-valued random variable. Many important properties of physical systems can be represented mathematically as matrix problems. For example, the thermal conductivity of a lattice can be computed from the dynamical matrix of the particle-particle interactions within the lattice.

Motivation

Physics

In nuclear physics, random matrices were introduced by Eugene Wigner to model the spectra of heavy atoms. He postulated that the spacings between the lines in the spectrum of a heavy atom should resemble the spacings between the eigenvalues of a random matrix, and should depend only on the symmetry class of the underlying evolution. In solid-state physics, random matrices model the behaviour of large disordered Hamiltonians in the mean field approximation.

In quantum chaos, the Bohigas–Giannoni–Schmit (BGS) conjecture asserts that the spectral statistics of quantum systems whose classical counterparts exhibit chaotic behaviour are described by random matrix theory. Random matrix theory has also found applications to quantum gravity in two dimensions, mesoscopic physics, and more

Mathematical Statistics and Numerical Analysis

In multivariate statistics, random matrices were introduced by John Wishart for statistical analysis of large samples. Significant results have been shown that extend the classical scalar Chernoff, Bernstein, and Hoeffding inequalities to the largest eigenvalues of finite sums of random Hermitian matrices. Corollary results are derived for the maximum singular values of rectangular matrices. In numerical analysis, random matrices have been used since the work of John von Neumann and Herman Goldstine to describe computation errors in operations such as matrix multiplication.

Number Theory

In number theory, the distribution of zeros of the Riemann zeta function (and other L-functions) is modelled by the distribution of eigenvalues of certain random matrices. The connection ©was first discovered by Hugh Montgomery and Freeman J. Dyson. It is connected to the Hilbert–Pσlya conjecture.

Gaussian Ensembles

The most studied random matrix ensembles are the Gaussian ensembles. The Gaussian unitary ensemble GUE(n) is described by the Gaussian measure with density

$$\frac{1}{Z_{\mathrm{GUE}(n)}} e^{-\frac{n}{2}\mathrm{tr}H^2}$$

on the space of $n \times n$ Hermitian matrices $H = (H_{ij})^n_{i,j=1}$. Here $Z_{\mathrm{GUE}}(n) = 2^{n/2}\,\pi^{n^2/2}$ is a normalization constant, chosen so that the integral of the density is equal to one. The term *unitary* refers to the fact that the distribution is invariant under unitary conjugation. The Gaussian unitary ensemble models Hamiltonians lacking time-reversal symmetry.

The **Gaussian orthogonal ensemble** GOE(n) is described by the Gaussian measure with density

$$\frac{1}{Z_{\mathrm{GOE}(n)}} e^{-\frac{n}{4}\mathrm{tr}H^2}$$

on the space of $n \times n$ real symmetric matrices $H = (H_{ij})n\ i,j=1$. Its distribution is invariant under orthogonal conjugation, and it models Hamiltonians with time-reversal symmetry. The Gaussian symplectic ensemble GSE(n) is described by the Gaussian measure with density

$$\frac{1}{Z_{\mathrm{GSE}(n)}} e^{-n\mathrm{tr}H^2}$$

on the space of $n \times n$ quaternionic Hermitian matrices $H = (H_{ij})^n_{i,j=1}$. Its distribution is invariant under conjugation by the symplectic group, and it models Hamiltonians with time-reversal symmetry but no rotational symmetry.

The joint probability density for the eigenvalues $\lambda_1, \lambda_2, \ldots, \lambda_n$ of GUE/GOE/GSE is given by

$$\frac{1}{Z_{\beta,n}} \prod_{k=1}^{n} e^{-\frac{\beta n}{4}\lambda_k^2} \prod_{i<j} \left|\lambda_j - \lambda_i\right|^{\beta}, \quad (1)$$

where $\beta = 1$ for GOE, $\beta = 2$ for GUE, and $\beta = 4$ for GSE; $Z_{\beta,n}$ is a normalisation constant which can be explicitly computed. In the case of GUE ($\beta = 2$), the formula (1) describes a determinantal point process.

Generalizations

Wigner matrices are random Hermitian matrices $H_n = (H_n(i,j))_{i,j=1}^n$ such that the entries

$$\{H_n(i,j), 1 \le i \le j \le n\}$$

above the main diagonal are independent random variables with zero mean, and

$$\{H_n(i,j), 1 \le i < j \le n\}$$

have identical second moments.

Invariant matrix ensembles are random Hermitian matrices with density on the space of real symmetric/ Hermitian/ quaternionic Hermitian matrices, which is of the form $\frac{1}{Z_n}e^{-n\mathrm{tr}V(H)}$, where the function V is called the potential. The Gaussian ensembles are the only common special cases of these two classes of random matrices.

Spectral Theory of Random Matrices

The spectral theory of random matrices studies the distribution of the eigenvalues as the size of the matrix goes to infinity.

Global Regime

In the *global regime*, one is interested in the distribution of linear statistics of the form $N_{f,H} = n^{-1}$ tr $f(H)$.

Empirical Spectral Measure

The *empirical spectral measure* μ_H of H is defined by

$$\mu_H(A) = \frac{1}{n}\#\{\text{eigenvalues of } H \text{ in } A\} = N_{1_A,H}, \quad A \subset \mathbb{R}.$$

Usually, the limit of μ_H is a deterministic measure; this is a particular case of self-averaging. The cumulative distribution function of the limiting measure is called the integrated density of states and is denoted $N(\lambda)$. If the integrated density of states is differentiable, its derivative is called the density of states and is denoted $\rho(\lambda)$.

The limit of the empirical spectral measure for Wigner matrices was described by Eugene Wigner. A more general theory was developed by Marchenko and Pastur. The limit of the empirical spectral measure of invariant matrix ensembles is described by a certain integral equation which arises from potential theory.

Fluctuations

For the linear statistics $N_{f,H} = n^{-1} - f(\lambda_j)$, one is also interested in the fluctuations about $+- f(\lambda) \, dN(\lambda)$. For many classes of random matrices, a central limit theorem of the form

$$\frac{N_{f,H} - \int f(\lambda)dN(\lambda)}{\sigma_{f,n}} \xrightarrow{D} N(0,1)$$

is known, etc.

Local Regime

In the *local regime*, one is interested in the spacings between eigenvalues, and, more generally, in the joint distribution of eigenvalues in an interval of length of order $1/n$.

One distinguishes between *bulk statistics*, pertaining to intervals inside the support of the limiting spectral measure, and *edge statistics*, pertaining to intervals near the boundary of the support.

Bulk Statistics

Formally, fix λ_0 in the interior of the support of $N(\lambda)$. Then consider the point process

$$\Xi(\lambda_0) = \sum_j \delta\left(\cdot - n\rho(\lambda_0)(\lambda_j - \lambda_0)\right),$$

where λ_j are the eigenvalues of the random matrix.

The point process $\Xi(\lambda_0)$ captures the statistical properties of eigenvalues in the vicinity of λ_0. For the Gaussian ensembles, the limit of $\Xi(\lambda_0)$ is known; thus, for GUE it is a determinantal point process with the kernel

$$K(x,y) = \frac{\sin \pi(x-y)}{\pi(x-y)}$$

(the *sine kernel*).

The *universality* principle postulates that the limit of $\Xi(\lambda_0)$ as $n \to \infty$ should depend only on the symmetry class of the random matrix (and neither on the specific model of random matrices nor on λ_0). This was rigorously proved for several models of random matrices: for invariant matrix ensembles, for Wigner matrices, et cet.

Other Classes of Random Matrices

Wishart Matrices

Wishart matrices are $n \times n$ random matrices of the form $H = XX^*$, where X is an $n \times n$ random matrix with independent entries, and X^* is its conjugate matrix. In the important special case considered by Wishart, the entries of X are identically distributed Gaussian random variables (either real or complex).

The limit of the empirical spectral measure of Wishart matrices was found by Vladimir Marchenko and Leonid Pastur.

Circular Ensemble

In the theory of random matrices, the circular ensembles are measures on spaces of unitary matrices introduced by Freeman Dyson as modifications of the Gaussian matrix ensembles. The three main examples are the orthogonal circular ensemble (COE) on symmetric unitary matrices, the unitary circular ensemble (CUE) on unitary matrices, and the symplectic circular ensemble (CSE) on self dual unitary quaternionic matrices.

The distribution of the unitary circular ensemble CUE(n) is the Haar measure on the unitary group $U(n)$. If U is a random element of CUE(n), then U^TU is a random element of COE(n); if U is a random element of CUE($2n$), then U^RU is a random element of CSE(n), where

$$U^R = \begin{pmatrix} 0 & -1 & & & & & \\ 1 & 0 & & & & & \\ & & 0 & -1 & & & \\ & & 1 & 0 & & & \\ & & & & \ddots & & \\ & & & & & 0 & -1 \\ & & & & & 1 & 0 \end{pmatrix} U^T \begin{pmatrix} 0 & 1 & & & & & \\ -1 & 0 & & & & & \\ & & 0 & 1 & & & \\ & & -1 & 0 & & & \\ & & & & \ddots & & \\ & & & & & 0 & 1 \\ & & & & & -1 & 0 \end{pmatrix}.$$

The joint distribution of eigenvalues of the circular ensembles is given by the probability density function

$$p(e^{i\theta_1},\cdots,e^{i\theta_n}) = \frac{1}{Z_{n,\beta}} \prod_{1\le k<j\le n} |e^{i\theta_k} - e^{i\theta_j}|^\beta ,$$

where β=1 for COE, β=2 for CUE, and β=4 for CSE. The normalisation constant $Z_{n,\beta}$ is given by

$$Z_{n,\beta} = (2\pi)^n \frac{\Gamma(\beta n/2+1)}{\left(\Gamma(\beta/2+1)\right)^n}.$$

Integrals of products of matrix coefficients in the unitary circular ensemble can be calculated using Weingarten functions.

Circular Law

In probability theory, more specifically the study of random matrices, the circular law describes the distribution of eigenvalues of an $n \times n$ random matrix with independent and identically distributed entries in the limit $n \to \infty$.

It asserts that for any sequence of random matrices whose entries are independent and identically distributed random variables, all with mean zero and variance equal to *1/n*, the limiting spectral distribution is the uniform distribution over the unit disc.

Precise Statement

Let $(X_n)_{n=1}^{\infty}$ be a sequence of $n \times n$ matrix ensembles whose entries are i.i.d. copies of a complex random variable x with mean 0 and variance 1. Let $\lambda_1, \ldots, \lambda_n, 1 \le j \le n$ denote the eigenvalues of $\frac{1}{\sqrt{n}} X_n$. Define the empirical spectral measure of $\frac{1}{\sqrt{n}} X_n$ as

$$\mu_{\frac{1}{\sqrt{n}}X_n}(A) = n^{-1} \#\{j \le n : \lambda_j \in A\}, \quad A \in \mathcal{B}(\mathbb{C})$$

With these definitions in mind, the circular law asserts the sequence $\mu_{\frac{1}{\sqrt{n}}X_n}(x,y)$ converges weakly in distribution to the uniform measure on the unit disk.

History

For random matrices with Gaussian distribution of entries (the *Ginibre ensembles*), the circular law was established in the 1960-s by Jean Ginibre. In the 1980s, Vyacheslav Girko introduced an approach which allowed to establish the circular law for more general distributions. Further progress was made by Zhidong Bai, who established the circular law under certain smoothness assumptions on the distribution.

The assumptions were further relaxed in the works of Terence Tao and Van H. Vu, Friedrich Götze and Alexander Tikhomirov. Finally, in 2010 Tao and Vu proved the circular law under the minimal assumptions stated above.

Tracy–Widom Distribution

The Tracy–Widom distribution, introduced by Craig Tracy and Harold Widom (1993, 1994), is the probability distribution of the largest eigenvalue of a random hermitian matrix in the edge scaling limit. It also appears in the distribution of the length of the longest increasing subsequence of random permutations (Baik, Deift & Johansson 1999) and in current fluctuations of the asymmetric simple exclusion process (ASEP) with step initial condition (Johansson 2000, Tracy & Widom 2009).

The cumulative distribution function of the Tracy–Widom distribution can be given as the Fredholm determinant

$$F_2(s) = \det(I - A_s)$$

of the operator A_s on square integrable function on the half line $(s, -)$ with kernel given in terms of Airy functions Ai by

$$\frac{\mathrm{Ai}(x)\mathrm{Ai}'(y) - \mathrm{Ai}'(x)\mathrm{Ai}(y)}{x - y}.$$

It can also be given as an integral

$$F_2(s) = \exp\left(-\int_s^{\infty}(x-s)q^2(x)dx\right)$$

in terms of a solution of a Painlevé equation of type II

$$q''(s) = sq(s) + 2q(s)^3$$

where q, called the Hastings-McLeod solution, satisfies the boundary condition

$$q(s) \sim \mathrm{Ai}(s), s \to \infty.$$

The distribution F_2 is associated to unitary ensembles in random matrix theory. There are analogous Tracy–Widom distributions F_1 and F_4 for orthogonal (β=1) and symplectic ensembles (β=4) that are also expressible in terms of the same Painlevé transcendent q (Tracy & Widom 1996):

$$F_1(s) = \exp\left(-\frac{1}{2}\int_s^\infty q(x)dx\right)\left(F_2(s)\right)^{1/2}$$

and

$$F_4(s/\sqrt{2}) = \cosh\left(\frac{1}{2}\int_s^\infty q(x)dx\right)\left(F_2(s)\right)^{1/2}.$$

The distribution F_1 is of particular interest in multivariate statistics (Johnstone 2007, 2008, 2009). For a discussion of the universality of F_β, β=1,2, and 4. For an application of F_1 to inferring population structure from genetic data, Price & Reich (2006).

Numerical techniques for obtaining numerical solutions to the Painlevé equations of the types II and V, and numerically evaluating eigenvalue distributions of random matrices in the beta-ensembles were first presented by Edelman & Persson (2005) using MATLAB. These approximation techniques were further analytically justified in Bejan (2005) and used to provide numerical evaluation of Painlevé II and Tracy–Widom distributions (for β=1,2, and 4) in S-PLUS. These distributions have been tabulated in Bejan (2005) to four significant digits for values of the argument in increments of 0.01; a statistical table for p-values was also given in this work. Bornemann (2009) gave accurate and fast algorithms for the numerical evaluation of F_β and the density functions $f_\beta(s)=dF_\beta/ds$ for β=1,2, and 4. These algorithms can be used to compute numerically the mean, variance, skewness and kurtosis of the distributions F_β.

β	*Mean*	*Variance*	*Skewness*	*Kurtosis*
1	-1.2065335745820	1.607781034581	0.29346452408	0.1652429384
2	-1.771086807411	0.8131947928329	0.224084203610	0.0934480876
4	-2.306884893241	0.5177237207726	0.16550949435	0.0491951565

Functions for working with the Tracy-Widom laws are also presented in the R package 'RMTstat' by Johnstone et al. (2009) and MATLAB package 'RMLab' by Dieng (2006).

For an extension of the definition of the Tracy–Widom distributions F_β to all β>0, Ramírez, Rider & Virág (2006).

Chapter 6

Diagonalizable Matrix

In linear algebra, a square matrix A is called diagonalizable if it is similar to a diagonal matrix, i.e., if there exists an invertible matrix P such that $P^{-1}AP$ is a diagonal matrix. If V is a finite-dimensional vector space, then a linear map $T : V \to V$ is called diagonalizable if there exists a basis of V with respect to T which is represented by a diagonal matrix. Diagonalization is the process of finding a corresponding diagonal matrix for a diagonalizable matrix or linear map. A square matrix which is not diagonalizable is called *defective*.

Diagonalizable matrices and maps are of interest because diagonal matrices are especially easy to handle: their eigenvalues and eigenvectors are known and one can raise a diagonal matrix to a power by simply raising the diagonal entries to that same power. Geometrically, a diagonalizable matrix is an inhomogeneous dilation (or *anisotropic scaling*) – it scales the space, as does a *homogeneous dilation*, but by a different factor in each direction, determined by the scale factors on each axis (diagonal entries).

Characterisation

The fundamental fact about diagonalizable maps and matrices is expressed by the following:

- An n-by-n matrix A over the field F is diagonalizable if and only if the sum of the dimensions of its eigenspaces is equal to n, which is the case if and only if there exists a basis of F^n consisting of eigenvectors of A. If such a basis has been found, one can form the matrix P having these basis vectors as columns, and $P^{-1}AP$ will be a diagonal matrix. The diagonal entries of this matrix are the eigenvalues of A.

- A linear map $T: V \to V$ is diagonalizable if and only if the sum of the dimensions of its eigenspaces is equal to $\dim(V)$, which is the case if and only if there exists a basis of V consisting of eigenvectors of T. With respect to such a basis, T will be represented by a diagonal matrix. The diagonal entries of this matrix are the eigenvalues of T.

Another characterization: A matrix or linear map is diagonalizable over the field F if and only if its minimal polynomial is a product of distinct linear factors over F. (Put in another way, a matrix is diagonalizable if and only if all of its elementary divisors are linear.)

The following sufficient (but not necessary) condition is often useful.

- An n-by-n matrix A is diagonalizable over the field F if it has n distinct eigenvalues in F, i.e. if its characteristic polynomial has n distinct roots in F; however, the converse may be false. Let us consider

$$\begin{bmatrix} -1 & 3 & -1 \\ -3 & 5 & -1 \\ -3 & 3 & 1 \end{bmatrix},$$

which has eigenvalues 1, 2, 2 (not all distinct) and is diagonalizable with diagonal form (also the similar matrix of A)

$$\begin{bmatrix} 1 & 0 & 0 \\ 0 & 2 & 0 \\ 0 & 0 & 2 \end{bmatrix}$$

and change of basis matrix P

$$\begin{bmatrix} 1 & 1 & -1 \\ 1 & 1 & 0 \\ 1 & 0 & 3 \end{bmatrix}.$$

- A linear map $T: V \to V$ with $n = \dim(V)$ is diagonalizable if it has n distinct eigenvalues, i.e. if its characteristic polynomial has n distinct roots in F.

Let A be a matrix over F. If A is diagonalizable, then so is any power of it. Conversely, if A is invertible, F is algebraically closed, and A^n is diagonalizable for some n that is not an integer multiple of the characteristic of F, then A is diagonalizable. Proof: If A^n is diagonalizable, then A is annihilated by some polynomial $(x^n - \lambda_1)\cdots(x^n - \lambda_k)$, which has no multiple root (since $\lambda_j \neq 0$) and is divided by the minimal polynomial of A.

As a rule of thumb, over C almost every matrix is diagonalizable. More precisely: the set of complex n-by-n matrices that are *not* diagonalizable over C, considered as a subset of $C^{n\times n}$, has Lebesgue measure zero.

One can also say that the diagonalizable matrices form a dense subset with respect to the Zariski topology: the complement lies inside the set where the discriminant of the characteristic polynomial vanishes, which is a hypersurface. From that follows also density in the usual (*strong*) topology given by a norm. The same is not true over R.

The Jordan–Chevalley decomposition expresses an operator as the sum of its semisimple (i.e., diagonalizable) part and its nilpotent part. Hence, a matrix is diagonalizable if and only if its nilpotent part is zero. Put in another way, a matrix is diagonalizable if each block in its Jordan form has no nilpotent part; i.e., one-by-one matrix.

Diagonalization

If a matrix A can be diagonalized, that is,

$$P^{-1}AP = \begin{pmatrix} \lambda_1 & & & \\ & \lambda_2 & & \\ & & \ddots & \\ & & & \lambda_n \end{pmatrix},$$

then:

$$AP = P\begin{pmatrix} \lambda_1 & & & \\ & \lambda_2 & & \\ & & \ddots & \\ & & & \lambda_n \end{pmatrix}.$$

Writing P as a block matrix of its column vectors

$$P = \left(\vec{\alpha}_1 \quad \vec{\alpha}_2 \quad \cdots \quad \vec{\alpha}_n\right),$$

the above equation can be rewritten as

$$A\vec{\alpha}_i = \lambda_i\vec{\alpha}_i \qquad (i = 1,2,\cdots,n).$$

So the column vectors of P are eigenvectors of A, and the corresponding diagonal entry is the corresponding eigenvalue. The invertibility of P also suggests that the eigenvectors are linearly independent and form the basis of F^n. This is the necessary and sufficient condition for diagonalizability and the canonical approach of diagonalization. When the matrix A is a Hermitian matrix (resp. symmetric matrix), eigenvectors of A can be chosen to form an orthonormal basis of C^n (resp. R^n). Under such circumstance P will be a unitary matrix (resp. orthogonal matrix) and P^{-1} equals the conjugate transpose (resp. transpose) of P.

Simultaneous Diagonalization

A set of matrices are said to be *simultaneously diagonalisable* if there exists a single invertible matrix P such that $P^{-1}AP$ is a diagonal matrix for every A in the set. The following theorem characterises simultaneously diagonalisable matrices: A set of diagonalizable matrices commutes if and only if the set is simultaneously diagonalisable.

The set of all n-by-n diagonalisable matrices (over C) with $n > 1$ is not simultaneously diagonalisable. For instance, the matrices

$$\begin{bmatrix} 1 & 0 \\ 0 & 0 \end{bmatrix} \text{ and } \begin{bmatrix} 1 & 1 \\ 0 & 0 \end{bmatrix}$$

are diagonalizable but not simultaneously diagonalizable because they do not commute.

A set consists of commuting normal matrices if and only if it is simultaneously diagonalisable by a unitary matrix; that is, there exists a unitary matrix U such that U^*AU is diagonal for every A in the set. In the language of Lie theory, a set of simultaneously diagonalisable matrices generate a toral Lie algebra.

Examples

Diagonalizable Matrices:

- Involutions are diagonalisable over the reals (and indeed any field of characteristic not 2), with 1's and -1's on the diagonal
- Finite order endomorphisms are diagonalisable over the complex numbers (or any algebraically closed field where the characteristic of the field does not divide the order of the endomorphism) with roots of unity on the diagonal. This follows since the minimal polynomial is separable, because the roots of unity are distinct.
- Projections are diagonalizable, with 0's and 1's on the diagonal.
- Real symmetric matrices are diagonalizable by orthogonal matrices; i.e., given a real symmetric matrix $A, Q^T AQ$ is diagonal for some orthogonal matrix Q.

  More generally, matrices are diagonalizable by unitary matrices if and only if they are normal.

  In the case of the real symmetric matrix, we see that $A = A^T$, so clearly $AA^T = A^T A$ holds. Examples of normal matrices are real symmetric (or skew-symmetric) matrices (e.g. covariance matrices) and Hermitian matrices (or skew-Hermitian matrices). See spectral theorems for generalizations to infinite-dimensional vector spaces.

Matrices That are not Diagonalizable

Some matrices are not diagonalizable over any field, most notably nonzero nilpotent matrices. This happens more generally if the algebraic and geometric multiplicities of an eigenvalue do not coincide. For instance, consider

$$C = \begin{bmatrix} 0 & 1 \\ 0 & 0 \end{bmatrix}.$$

This matrix is not diagonalizable: there is no matrix U such that $U^{-1}CU$ is a diagonal matrix. Indeed, C has one eigenvalue (namely zero) and this eigenvalue has algebraic multiplicity 2 and geometric multiplicity 1.

Some real matrices are not diagonalizable over the reals. Consider for instance the matrix

$$B = \begin{bmatrix} 0 & 1 \\ -1 & 0 \end{bmatrix}.$$

The matrix B does not have any real eigenvalues, so there is no real matrix Q such that $Q^{-1}BQ$ is a diagonal matrix. However, we can diagonalize B if we allow complex numbers. Indeed, if we take

$$Q = \begin{bmatrix} 1 & i \\ i & 1 \end{bmatrix},$$

then $Q^{-1}BQ$ is diagonal.

Note that the above examples show that the sum of diagonalizable matrices need not be diagonalizable.

Quantum Mechanical Application

In quantum mechanical and quantum chemical computations matrix diagonalization is one of the most frequently applied numerical processes. The basic reason is that the time-independent Schrödinger equation is an eigenvalue equation, albeit in most of the physical situations on an infinite dimensional space (a Hilbert space).

A very common approximation is to truncate Hilbert space to finite dimension, after which the Schrödinger equation can be formulated as an eigenvalue problem of a real symmetric, or complex Hermitian, matrix. Formally this approximation is founded on the variational principle, valid for Hamiltonians that are bounded from below. But also first-order perturbation theory for degenerate states leads to a matrix eigenvalue problem.

Gaussian Elimination

In linear algebra, Gaussian elimination is an algorithm for solving systems of linear equations. It can also be used to find the rank of a matrix, to calculate the determinant of a matrix, and to calculate the inverse of an invertible square matrix. The method is named after Carl Friedrich Gauss, but it was not invented by him.

Elementary row operations are used to reduce a matrix to what is called triangular form (in numerical analysis) or row echelon form (in abstract algebra). Gauss–Jordan elimination, an extension of this algorithm, reduces the matrix further to diagonal form, which is also known as reduced row echelon form. Gaussian elimination alone is sufficient for many applications, and requires fewer calculations than the Gauss–Jordan version.

History

The method of Gaussian elimination appears in the important Chinese mathematical text Chapter Eight *Rectangular Arrays* of *The Nine Chapters on the Mathematical Art*. Its use is illustrated in eighteen problems, with two to five equations. The first reference to the book by this title is dated to 179 CE, but parts of it were written as early as approximately 150 BCE. It was commented on by Liu Hui in the 3rd century.

The method in Europe stems from the notes of Isaac Newton. In 1670, he wrote that all the algebra books known to him lacked a lesson for solving simultaneous equations, which Newton then supplied. Cambridge University eventually published the notes as *Arithmetica Universalis* in 1707 long after Newton left academic life. The notes were widely imitated, which made (what is now called) Gaussian elimination a standard lesson in algebra textbooks by the end of the 18th century. Carl Friedrich Gauss in 1810 devised a notation for symmetric elimination that was adopted in the 19th century by professional hand computers to solve the normal equations of least-squares problems. The algorithm that is taught in high school was named for Gauss only in the 1950s as a result of confusion over the history of the subject.

Algorithm Overview

The process of Gaussian elimination has two parts. The first part (Forward Elimination) reduces a given system to either triangular or echelon form, or results in a degenerate equation, indicating the system has no unique solution but may have multiple solutions (rank<order). This is accomplished through the use of elementary row operations. The second step uses back substitution to find the solution of the system above.

Stated equivalently for matrices, the first part reduces a matrix to row echelon form using elementary row operations while the second reduces it to reduced row echelon form, or row canonical form.

Another point of view, which turns out to be very useful to analyze the algorithm, is that Gaussian elimination computes a matrix decomposition. The three elementary row operations used in the Gaussian elimination (multiplying rows, switching rows, and adding multiples of rows to other rows) amount to multiplying the original matrix with invertible matrices from the left.

The first part of the algorithm computes an LU decomposition, while the second part writes the original matrix as the product of a uniquely determined invertible matrix and a uniquely determined reduced row-echelon matrix.

Example

Suppose the goal is to find and describe the solution(s), if any, of the following system of linear equations:

$$\begin{array}{rcrcrcrl} 2x & + & y & - & z & = & 8 & (L_1) \\ -3x & - & y & + & 2z & = & -11 & (L_2) \\ -2x & + & y & + & 2z & = & -3 & (L_3) \end{array}$$

The algorithm is as follows: eliminate x from all equations below L_1, and then eliminate y from all equations below L_2. This will put the system into triangular form. Then, using back-substitution, each unknown can be solved for.

In the example, x is eliminated from L_2 by adding $\frac{3}{2}L_1$ to L_2. x is then eliminated from L_3 by adding L_1 to L_3. Formally:

$$L_2 + \frac{3}{2}L_1 \rightarrow L_2$$

$$L_3 + L_1 \rightarrow L_3$$

The result is:

$$\begin{array}{rcrcrcr} 2x & + & y & - & z & = & 8 \\ & & \frac{1}{2}y & + & \frac{1}{2}z & = & 1 \\ & & 2y & + & z & = & 5 \end{array}$$

Now y is eliminated from L_3 by adding $-4L_2$ to: L_3

$$L_3 + -4L_2 \rightarrow L_3$$

The result is:

$$\begin{array}{ccccccc} 2x & + & y & - & z & = & 8 \\ & & \frac{1}{2}y & + & \frac{1}{2}z & = & 1 \\ & & & & -z & = & 1 \end{array}$$

This result is a system of linear equations in triangular form, and so the first part of the algorithm is complete.

The last part, back-substitution, consists of solving for the knowns in reverse order. It can thus be seen that

$$z = -1 \quad (L_3)$$

Then, z can be substituted into L_2, which can then be solved to obtain

$$y = 3 \quad (L_2)$$

Next, z and y can be substituted into L_1, which can be solved to obtain

$$x = 2 \quad (L_1)$$

The system is solved.

Some systems cannot be reduced to triangular form, yet still have at least one valid solution: for example, if y had not occurred in L_2 and L_3 after the first step above, the algorithm would have been unable to reduce the system to triangular form.

However, it would still have reduced the system to echelon form. In this case, the system does not have a unique solution, as it contains at least one free variable.

The solution set can then be expressed parametrically (that is, in terms of the free variables, so that if values for the free variables are chosen, a solution will be generated).

In practice, one does not usually deal with the systems in terms of equations but instead makes use of the augmented matrix (which is also suitable for computer manipulations). For example:

$$\begin{array}{ccccccccc} 2x & + & y & - & z & = & 8 & & (L_1) \\ -3x & - & y & + & 2z & = & -11 & & (L_2) \\ -2x & + & y & + & 2z & = & -3 & & (L_3) \end{array}$$

Therefore, the Gaussian Elimination algorithm applied to the augmented matrix begins with:

$$\begin{bmatrix} 2 & 1 & -1 & 8 \\ -3 & -1 & 2 & -11 \\ -2 & 1 & 2 & -3 \end{bmatrix}$$

which, at the end of the first part (Gaussian elimination, zeros only under the leading 1) of the algorithm, looks like this:

$$\begin{bmatrix} 1 & \frac{1}{3} & \frac{-2}{3} & \frac{11}{3} \\ 0 & 1 & \frac{2}{5} & \frac{13}{5} \\ 0 & 0 & 1 & -1 \end{bmatrix}$$

That is, it is in row echelon form.

At the end of the algorithm, if the Gauss–Jordan elimination(zeros under and above the leading 1) is applied:

$$\begin{bmatrix} 1 & 0 & 0 & 2 \\ 0 & 1 & 0 & 3 \\ 0 & 0 & 1 & -1 \end{bmatrix}$$

That is, it is in reduced row echelon form, or row canonical form.

Chapter 7

Band Matrix

In mathematics, particularly matrix theory, a band matrix is a sparse matrix whose non-zero entries are confined to a diagonal *band*, comprising the main diagonal and zero or more diagonals on either side.

Matrix Bandwidth

Formally, consider an $n \times n$ matrix $A=(a_{i,j})$. If all matrix elements are zero outside a diagonally bordered band whose range is determined by constants k_1 and k_2:

$$a_{i,j} = 0 \quad \text{if} \quad j < i - k_1 \quad \text{or} \quad j > i + k_2; \quad k_1, k_2 \geq 0.$$

then the quantities k_1 and k_2 are called the left and right half-bandwidth, respectively (Golub & Van Loan 1996, §1.2.1). The bandwidth of the matrix is $k_1 + k_2 + 1$ (in other words, it is the smallest number of adjacent diagonals to which the non-zero elements are confined).

A matrix is called a band matrix or banded matrix if its bandwidth is reasonably small.

A band matrix with $k_1 = k_2 = 0$ is a diagonal matrix; a band matrix with $k_1 = k_2 = 1$ is a tridiagonal matrix; when $k_1 = k_2 = 2$ one has a pentadiagonal matrix and so on. If one puts $k_1 = 0$, $k_2 = n-1$, one obtains the definition of an upper triangular matrix; similarly, for $k_1 = n-1$, $k_2 = 0$ one obtains a lower triangular matrix.

Applications

In numerical analysis, matrices from finite element or finite difference problems are often banded.

Such matrices can be viewed as descriptions of the coupling between the problem variables; the bandedness corresponds to the fact that variables are not coupled over arbitrarily large distances. Such matrices can be further divided - for instance, banded matrices exist where every element in the band is nonzero. These often arise when discretising one-dimensional problems.

Problems in higher dimensions also lead to banded matrices, in which case the band itself also tends to be sparse.

For instance, a partial differential equation on a square domain (using central differences) will yield a matrix with a half-bandwidth equal to the square root of the matrix dimension, but inside the band only 5 diagonals are nonzero.

Unfortunately, applying Gaussian elimination (or equivalently an LU decomposition) to such a matrix results in the band being filled in by many non-zero elements.

Band Storage

Band matrices are usually stored by storing the diagonals in the band; the rest is implicitly zero.

For example, a tridiagonal matrix has bandwidth 1. The 6-by-6 matrix

$$\begin{bmatrix} B_{11} & B_{12} & 0 & \cdots & \cdots & 0 \\ B_{21} & B_{22} & B_{23} & \ddots & \ddots & \vdots \\ 0 & B_{32} & B_{33} & B_{34} & \ddots & \vdots \\ \vdots & \ddots & B_{43} & B_{44} & B_{45} & 0 \\ \vdots & \ddots & \ddots & B_{54} & B_{55} & B_{56} \\ 0 & \cdots & \cdots & 0 & B_{65} & B_{66} \end{bmatrix}$$

is stored as the 6-by-3 matrix

$$\begin{bmatrix} 0 & B_{11} & B_{12} \\ B_{21} & B_{22} & B_{23} \\ B_{32} & B_{33} & B_{34} \\ B_{43} & B_{44} & B_{45} \\ B_{54} & B_{55} & B_{56} \\ B_{65} & B_{66} & 0 \end{bmatrix}.$$

A further saving is possible when the matrix is symmetric. For example, consider a symmetric 6-by-6 matrix with a right bandwidth of 2:

$$\begin{bmatrix} A_{11} & A_{12} & A_{13} & 0 & \cdots & 0 \\ & A_{22} & A_{23} & A_{24} & \ddots & \vdots \\ & & A_{33} & A_{34} & A_{35} & 0 \\ & & & A_{44} & A_{45} & A_{46} \\ & sym & & & A_{55} & A_{56} \\ & & & & & A_{66} \end{bmatrix}.$$

This matrix is stored as the 6-by-3 matrix:

$$\begin{bmatrix} A_{11} & A_{12} & A_{13} \\ A_{22} & A_{23} & A_{24} \\ A_{33} & A_{34} & A_{35} \\ A_{44} & A_{45} & A_{46} \\ A_{55} & A_{56} & 0 \\ A_{66} & 0 & 0 \end{bmatrix}.$$

Band Form of Sparse Matrices

From a computational point of view, working with band matrices is always preferential to working with similarly dimensioned square matrices. A band matrix can be likened in complexity to a rectangular matrix whose row dimension is equal to the bandwidth of the band matrix. Thus the work involved in performing operations such as multiplication falls significantly, often leading to huge savings in terms of calculation time and complexity.

As sparse matrices lend themselves to more efficient computation than dense matrices, as well as in more efficient utilization of computer storage, there has been much research focused on finding ways to minimise the bandwidth (or directly minimise the fill in) by applying permutations to the matrix, or other such equivalence or similarity transformations. The Cuthill–McKee algorithm can be used to reduce the bandwidth of a sparse symmetric matrix. There are, however, matrices for which the reverse Cuthill–McKee algorithm performs better. There are many other methods in use. The problem of finding a representation of a matrix with minimal bandwidth by means of permutations of rows and columns is NP-hard.

Examples and Special Cases

The following are special cases of band matrices:

- Diagonal matrices.
- Tridiagonal matrices.
- Pentadiagonal matrices.
- Upper and lower triangular matrices.
- Upper and lower Hessenberg matrices.
- Block-diagonal matrices.
- Shift matrices and shear matrices.
- Matrices in Jordan normal form.
- A skyline matrix, also called "variable band matrix" is a generalization of band matrix

The inverses of Lehmer matrices are constant tridiagonal matrices, and are thus band matrices.

Bidiagonal Matrix

In mathematics, a bidiagonal matrix is a matrix with non-zero entries along the main diagonal and *either* the diagonal above or the diagonal below. This means there are exactly two non zero diagonals in the matrix. When the diagonal above the main diagonal has the non-zero entries the matrix is upper bidiagonal. When the diagonal below the main diagonal has the non-zero entries the matrix is lower bidiagonal. For example, the following matrix is upper bidiagonal:

$$\begin{pmatrix} 1 & 4 & 0 & 0 \\ 0 & 4 & 1 & 0 \\ 0 & 0 & 3 & 4 \\ 0 & 0 & 0 & 3 \end{pmatrix}$$

and the following matrix is lower bidiagonal:

$$\begin{pmatrix} 1 & 0 & 0 & 0 \\ 2 & 4 & 0 & 0 \\ 0 & 3 & 3 & 0 \\ 0 & 0 & 4 & 3 \end{pmatrix}.$$

Usage

One variant of the QR algorithm starts with reducing a general matrix into a bidiagonal one. and the Singular value decomposition uses this method as well.

Block Matrix

In the mathematical discipline of matrix theory, a block matrix or a partitioned matrix is a matrix broken into sections called blocks. Looking at it another way, the matrix is written in terms of smaller matrices. We group the rows and columns into adjacent 'bunches'.

A partition is the rectangle described by one bunch of adjacent rows and one bunch of adjacent columns. Rows and columns must be formed consistently:

the matrix is split into blocks by horizontal and vertical lines, which must cut the matrix completely in the given direction.

The block matrix may also have an inverse.

Example

The matrix

$$\mathbf{P} = \begin{bmatrix} 1 & 1 & 2 & 2 \\ 1 & 1 & 2 & 2 \\ 3 & 3 & 4 & 4 \\ 3 & 3 & 4 & 4 \end{bmatrix}$$

can be partitioned into 4 2×2 blocks

$$\mathbf{P}_{11} = \begin{bmatrix} 1 & 1 \\ 1 & 1 \end{bmatrix}, \mathbf{P}_{12} = \begin{bmatrix} 2 & 2 \\ 2 & 2 \end{bmatrix}, \mathbf{P}_{21} = \begin{bmatrix} 3 & 3 \\ 3 & 3 \end{bmatrix}, \mathbf{P}_{22} = \begin{bmatrix} 4 & 4 \\ 4 & 4 \end{bmatrix}.$$

The partitioned matrix can then be written as

$$\mathbf{P} = \begin{bmatrix} \mathbf{P}_{11} & \mathbf{P}_{12} \\ \mathbf{P}_{21} & \mathbf{P}_{22} \end{bmatrix}.$$

Block Matrix Multiplication

A block partitioned matrix product can be formed involving operations only on the submatrices. Given an $(m \times p)$ matrix $A$with q row partitions and $s$column partitions

$$\mathbf{A} = \begin{bmatrix} \mathbf{A}_{11} & \mathbf{A}_{12} & \cdots & \mathbf{A}_{1s} \\ \mathbf{A}_{21} & \mathbf{A}_{22} & \cdots & \mathbf{A}_{2s} \\ \vdots & \vdots & \ddots & \vdots \\ \mathbf{A}_{q1} & \mathbf{A}_{q2} & \cdots & \mathbf{A}_{qs} \end{bmatrix}$$

and a $(m \times p)$ matrix B with row partitions and column partitions

$$\mathbf{B} = \begin{bmatrix} \mathbf{B}_{11} & \mathbf{B}_{12} & \cdots & \mathbf{B}_{1r} \\ \mathbf{B}_{21} & \mathbf{B}_{22} & \cdots & \mathbf{B}_{2r} \\ \vdots & \vdots & \ddots & \vdots \\ \mathbf{B}_{s1} & \mathbf{B}_{s2} & \cdots & \mathbf{B}_{sr} \end{bmatrix},$$

the matrix product

$$\mathbf{C} = \mathbf{AB}$$

can be formed blockwise, yielding C as an $(m \times n)$ matrix with q row partitions and r column partitions. The matrices in your matrix are calculated by multiplying while you multiply:

$$\mathbf{C}_{\alpha\beta} = \sum_{\gamma=1}^{s} \mathbf{A}_{\alpha\gamma}\mathbf{B}_{\gamma\beta}.$$

Block Diagonal Matrices

A block diagonal matrix is a block matrix which is a square matrix, and having main diagonal blocks square matrices, such that the off-diagonal blocks are zero matrices. A block diagonal matrix A has the form

$$\mathbf{A} = \begin{bmatrix} \mathbf{A}_1 & 0 & \cdots & 0 \\ 0 & \mathbf{A}_2 & \cdots & 0 \\ \vdots & \vdots & \ddots & \vdots \\ 0 & 0 & \cdots & \mathbf{A}_n \end{bmatrix}$$

where A_k is a square matrix; in other words, it is the direct sum of A_1, ..., A_n. It can also be indicated as $\mathbf{A}_1 \oplus \mathbf{A}_2 \oplus \ldots \oplus \mathbf{A}_n$ or diag(A_1, A_2,..., A_n) (the latter being the same formalism used for a diagonal matrix). Any square matrix can trivially be considered a block diagonal matrix with only one block.

For the determinant and trace, the following properties hold

$$\det \mathbf{A} = \det \mathbf{A}_1 \cdot \ldots \cdot \det \mathbf{A}_n,$$

$$\operatorname{trace} \mathbf{A} = \operatorname{trace} \mathbf{A}_1 + \cdots + \operatorname{trace} \mathbf{A}_n.$$

The inverse of a block diagonal matrix is another block diagonal matrix, composed of the inverse of each block, as follow :

$$\begin{pmatrix} \mathbf{A}_1 & 0 & \cdots & 0 \\ 0 & \mathbf{A}_2 & \cdots & 0 \\ \vdots & \vdots & \ddots & \vdots \\ 0 & 0 & \cdots & \mathbf{A}_n \end{pmatrix}^{-1} = \begin{pmatrix} \mathbf{A}_1^{-1} & 0 & \cdots & 0 \\ 0 & \mathbf{A}_2^{-1} & \cdots & 0 \\ \vdots & \vdots & \ddots & \vdots \\ 0 & 0 & \cdots & \mathbf{A}_n^{-1} \end{pmatrix}.$$

The eigenvalues and eigenvectors of $A$are simply those of A_1 and A_2 and ... and A_n(combined).

Block Tridiagonal Matrices

A block tridiagonal matrix is another special block matrix, which is just like the block diagonal matrix a square matrix, having square matrices (blocks) in the lower diagonal, main diagonal and upper diagonal, with all other blocks being zero matrices. It is essentially a tridiagonal matrix but has submatrices in places of scalars. A block tridiagonal matrix A has the form

$$\mathbf{A} = \begin{bmatrix} \mathbf{B}_1 & \mathbf{C}_1 & & & \cdots & & 0 \\ \mathbf{A}_2 & \mathbf{B}_2 & \mathbf{C}_2 & & & & \\ & \ddots & \ddots & \ddots & & & \vdots \\ & & \mathbf{A}_k & \mathbf{B}_k & \mathbf{C}_k & & \\ \vdots & & & \ddots & \ddots & \ddots & \\ & & & & \mathbf{A}_{n-1} & \mathbf{B}_{n-1} & \mathbf{C}_{n-1} \\ 0 & & \cdots & & & \mathbf{A}_n & \mathbf{B}_n \end{bmatrix}$$

where A_k, B_k and C_k are square sub-matrices of the lower, main and upper diagonal respectively. Block tridiagonal matrices are often encountered in numerical solutions of engineering problems (e.g., computational fluid dynamics). Optimized numerical methods for LU factorization are available and hence efficient solution algorithms for equation systems with a block tridiagonal matrix as coefficient matrix. The Thomas algorithm, used for efficient solution of equation systems involving a tridiagonal matrix can also be applied using matrix operations to block tridiagonal matrices.

Block Toeplitz Matrices

A block Toeplitz matrix is another special block matrix, which contains blocks that are repeated down the diagonals of the matrix, as a Toeplitz matrix has elements repeated down the diagonal.

A block Toeplitz matrix A has the form

$$\mathbf{A} = \begin{bmatrix} \mathbf{A}_{(1,1)} & \mathbf{A}_{(1,2)} & & & & \cdots & \mathbf{A}_{(1,n-1)} & \mathbf{A}_{(1,n)} \\ \mathbf{A}_{(2,1)} & \mathbf{A}_{(1,1)} & \mathbf{A}_{(1,2)} & & & & & \mathbf{A}_{(1,n-1)} \\ & \ddots & \ddots & \ddots & & & & \vdots \\ & & \mathbf{A}_{(2,1)} & \mathbf{A}_{(1,1)} & \mathbf{A}_{(1,2)} & & & \\ \vdots & & & \ddots & \ddots & \ddots & & \\ \mathbf{A}_{(n-1,1)} & & & & \mathbf{A}_{(2,1)} & \mathbf{A}_{(1,1)} & & \mathbf{A}_{(1,2)} \\ \mathbf{A}_{(n,1)} & \mathbf{A}_{(n-1,1)} & \cdots & & & \mathbf{A}_{(2,1)} & & \mathbf{A}_{(1,1)} \end{bmatrix}.$$

Direct Sum

For any arbitrary matrices A (of size $m \times n$) and B (of size $p \times q$), we have the direct sum of A and B, denoted by A $\oplus$ B and defined as

$$\mathbf{A} \oplus \mathbf{B} = \begin{bmatrix} a_{11} & \cdots & a_{1n} & 0 & \cdots & 0 \\ \vdots & \cdots & \vdots & \vdots & \cdots & \vdots \\ a_{m1} & \cdots & a_{mn} & 0 & \cdots & 0 \\ 0 & \cdots & 0 & b_{11} & \cdots & b_{1q} \\ \vdots & \cdots & \vdots & \vdots & \cdots & \vdots \\ 0 & \cdots & 0 & b_{p1} & \cdots & b_{pq} \end{bmatrix}.$$

For instance,

$$\begin{bmatrix} 1 & 3 & 2 \\ 2 & 3 & 1 \end{bmatrix} \oplus \begin{bmatrix} 1 & 6 \\ 0 & 1 \end{bmatrix} = \begin{bmatrix} 1 & 3 & 2 & 0 & 0 \\ 2 & 3 & 1 & 0 & 0 \\ 0 & 0 & 0 & 1 & 6 \\ 0 & 0 & 0 & 0 & 1 \end{bmatrix}.$$

This operation generalizes naturally to arbitrary dimensioned arrays (provided that A and B have the same number of dimensions).

Direct Product

Application

In linear algebra terms, the use of a block matrix corresponds to having a linear mapping thought of in terms of corresponding 'bunches' of basis vectors. That again matches the idea of having distinguished direct sum decompositions of the domain and range. It is always particularly significant if a block is the zero matrix; that carries the information that a summand maps into a sub-sum.

Given the interpretation *via* linear mappings and direct sums, there is a special type of block matrix that occurs for square matrices (the case $m = n$). For those we can assume an interpretation as an endomorphism of an n-dimensional space V; the block structure in which the bunching of rows and columns is the same is of importance because it corresponds to having a single direct sum decomposition on V (rather than two). In that case, for example, the diagonal blocks in the obvious sense are all square. This type of structure is required to describe the Jordan normal form.

This technique is used to cut down calculations of matrices, column-row expansions, and many computer science applications, including VLSI chip design. An example is the Strassen algorithm for fast matrix multiplication, as well as the Hamming(7,4) encoding for error detection and recovery in data transmissions.

Diagonal Matrix

In linear algebra, a diagonal matrix is a matrix (usually a square matrix) in which the entries outside the main diagonal (↘) are all zero. The diagonal entries themselves may or may not be zero. Thus, the matrix D = ($d_{i,j}$) with n columns and n rows is diagonal if:

$$d_{i,j} = 0 \text{ if } i \neq j \qquad \forall i, j \in \{1, 2, \ldots, n\}.$$

For example, the following matrix is diagonal:

$$\begin{bmatrix} 1 & 0 & 0 \\ 0 & 4 & 0 \\ 0 & 0 & -3 \end{bmatrix}$$

The term *diagonal matrix* may sometimes refer to a rectangular diagonal matrix, which is an m-by-n matrix with only the entries of the form $d_{i,i}$ possibly non-zero. For example:

$$\begin{bmatrix} 1 & 0 & 0 \\ 0 & 4 & 0 \\ 0 & 0 & -3 \\ 0 & 0 & 0 \end{bmatrix} \text{or} \begin{bmatrix} 1 & 0 & 0 & 0 & 0 \\ 0 & 4 & 0 & 0 & 0 \\ 0 & 0 & -3 & 0 & 0 \end{bmatrix}$$

However, in the remainder of this article we will consider only square matrices. Any square diagonal matrix is also a symmetric matrix. Also, if the entries come from the field R or C, then it is a normal matrix as well. Equivalently, we can define a diagonal matrix as a matrix that is both upper- and lower-triangular. The identity matrix I_n and any square zero matrix are diagonal. A one-dimensional matrix is always diagonal.

Scalar Matrix

A diagonal matrix with all its main diagonal entries equal is a scalar matrix, that is, a scalar multiple λI of the identity matrix I. Its effect on a vector is scalar multiplication by λ. For example, a 3×3 scalar matrix has the form:

$$\begin{bmatrix} \lambda & 0 & 0 \\ 0 & \lambda & 0 \\ 0 & 0 & \lambda \end{bmatrix}.$$

The scalar matrices are the centre of the algebra of matrices: that is, they are precisely the matrices that commute with all other square matrices of the same size.

For an abstract vector space V (rather than the concrete vector space K), or more generally a module M over a ring R, with the endomorphism algebra End(M) (algebra of linear operators on M) replacing the algebra of matrices, the analog of scalar matrices are scalar transformations. Formally, scalar multiplication is a linear map, inducing a map $R \to \text{End}(M)$, (send a scalar λ to the corresponding scalar transformation, multiplication by λ) exhibiting End(M) as a R-algebra. For vector spaces, or more generally free modules $M \cong R^n$, for which the endomorphism algebra is isomorphic to a matrix algebra, the scalar transforms are exactly the centre of the endomorphism algebra, and similarly invertible transforms are the centre of the general linear group GL(V), where they are denoted by Z(V), follow the usual notation for the centre.

Matrix Operations

The operations of matrix addition and matrix multiplication are especially simple for diagonal matrices. Write $\text{diag}(a_1,...,a_n)$ for a diagonal matrix whose diagonal entries starting in the upper left corner are $a_1,...,a_n$. Then, for addition, we have

$$\text{diag}(a_1,...,a_n) + \text{diag}(b_1,...,b_n) = \text{diag}(a_1+b_1,...,a_n+b_n)$$

and for matrix multiplication,

$$\text{diag}(a_1,...,a_n) \cdot \text{diag}(b_1,...,b_n) = \text{diag}(a_1b_1,...,a_nb_n).$$

The diagonal matrix $\text{diag}(a_1,...,a_n)$ is invertible if and only if the entries $a_1,...,a_n$ are all non-zero. In this case, we have

$$\text{diag}(a_1,...,a_n)^{-1} = \text{diag}(a_1^{-1},...,a_n^{-1}).$$

In particular, the diagonal matrices form a subring of the ring of all n-by-n matrices.

Multiplying an n-by-n matrix A from the *left* with $\text{diag}(a_1,...,a_n)$ amounts to multiplying the i-th *row* of A by a_i for all i; multiplying the matrix A from the *right* with $\text{diag}(a_1,...,a_n)$ amounts to multiplying the i-th *column* of A by a_i for all i.

Other Properties

The eigenvalues of diag $(a_1, ..., a_n)$ are $a_1, ..., a_n$ with associated eigenvectors of $e_1, ..., e_n$, where the vector e_i is all zeros except a one in the ith row. The determinant of $\text{diag}(a_1, ..., a_n)$ is the product $a_1...a_n$.

The adjugate of a diagonal matrix is again diagonal.

A square matrix is diagonal if and only if it is triangular and normal.

Uses

Diagonal matrices occur in many areas of linear algebra. Because of the simple description of the matrix operation and eigenvalues/eigenvectors given above, it is always desirable to represent a given matrix or linear map by a diagonal matrix.

In fact, a given n-by-n matrix A is similar to a diagonal matrix (meaning that there is a matrix X such that $X^{-1}AX$ is diagonal) if and only if it has n linearly independent eigenvectors. Such matrices are said to be diagonalizable.

Over the field of real or complex numbers, more is true. The spectral theorem says that every normal matrix is unitarily similar to a diagonal matrix (if $AA^* = A^*A$ then there exists a unitary matrix U such that UAU^* is diagonal). Furthermore, the singular value decomposition implies that for any matrix A, there exist unitary matrices U and V such that UAV^* is diagonal with positive entries.

Operator Theory

In operator theory, particularly the study of PDEs, operators are particularly easy to understand and PDEs easy to solve if the operator is diagonal with respect to the basis with which one is working; this corresponds to a separable partial differential equation.

Therefore, a key technique to understanding operators is a change of coordinates – in the language of operators, an integral transform – which changes the basis to an eigenbasis of eigenfunctions: which makes the equation separable. An important example of this is the Fourier transform, which diagonalizes constant coefficient differentiation operators (or more generally translation invariant operators), such as the Laplacian operator, say, in the heat equation.

Especially easy are multiplication operators, which are defined as multiplication by (the values of) a fixed function – the values of the function at each point correspond to the diagonal entries of a matrix.

Chapter 8

Distance Matrix

In mathematics, computer science and graph theory, a distance matrix is a matrix (two-dimensional array) containing the distances, taken pairwise, of a set of points. This matrix will have a size of $N \times N$ where N is the number of points, nodes or vertices (often in a graph).

Comparison with Related Matrices

Comparison with Adjacency Matrix

Distance matrices are related to adjacency matrices, with the differences that (a) the latter only provides the information which vertices are connected but does not tell about *costs* or *distances* between the vertices and (b) an entry of a distance matrix is smaller if two elements are closer, while "close" (connected) vertices yield larger entries in an adjacency matrix.

Comparison with Euclidean Distance Matrix

Unlike an Euclidean distance matrix, the matrix does not need to be symmetric — that is, the values $x_{i,j}$ do not necessarily equal $x_{j,i}$. Similarly, the matrix values are not restricted to non-negative reals (as they would be in the Euclidean distance matrix) but rather can have negative values, zeros or imaginary numbers depending on the cost metric and specific use. Although it is often the case, distance matrices are not restricted to being hollow — that is, they can have non-zero entries on the main diagonal.

Doubly Stochastic Matrix

In mathematics, especially in probability and combinatorics, a doubly stochastic matrix (also called bistochastic), is a square matrix of nonnegative real numbers, each of whose rows and columns sum to 1. Thus, a doubly stochastic matrix is both left stochastic and right stochastic.

Such a transition matrix is necessarily a square matrix: if every row sums to one then the sum of all entries in the matrix must be equal to the number of rows, and since the same holds for columns, the number of rows and columns must be equal.

The class of $n \times n$ doubly stochastic matrices is a convex polytope in R^N (where $N = n^2$), known as the Birkhoff polytope, B_n. Its dimension is $(n - 1)^2$, since the row and column sums being equal to 1 imposes $2n - 1$ linear constraints (not $2n$, because if the total of all n columns is n, the same must be true of the total of all n rows).

The principal fact about doubly stochastic matrices is the Birkhoff–von Neumann theorem. This states that the set B_n of doubly stochastic matrices of order n is the convex hull of the set of permutation matrices (of order n), and furthermore that the vertices (extreme points) of B_n are precisely the permutation matrices.

Sinkhorn's theorem states that any matrix with strictly positive entries can be made doubly stochastic by pre- and post-multiplication by diagonal matrices.

For $n = 2$, all bistochastic matrices are unistochastic and orthostochastic, but for larger n it is not the case.

Down-Shift Operator

In mathematics, a down-shift operator is a binary matrix with ones only on the superdiagonal or subdiagonal, and zeroes elsewhere.

Edmonds Matrix

In graph theory, the Edmonds matrix A of a balanced bipartite graph $G(U,V,E)$ with sets of vertices $U = \{u_1, u_2, \ldots, u_n\}$ and $V = \{v_1, v_2, \ldots, v_n\}$ is defined by

$$A_{ij} = \begin{cases} x_{ij} & (u_i, v_j) \in E \\ 0 & (u_i, v_j) \notin E \end{cases}$$

where the x_{ij} are indeterminates. One application of the Edmonds matrix of a bipartite graph is that the graph admits a perfect matching if and only if the polynomial $\det(A_{ij})$ in the x_{ij} is not identically zero. Furthermore, the number of perfect matchings is equal to the number of monomials in the polynomial $\det(A)$, and is also equal to the permanent of A. The Edmonds matrix is named after Jack Edmonds. The Tutte matrix is a generalisation to non-bipartite graphs.

Euclidean Distance Matrix

In mathematics, a Euclidean distance matrix is an $n \times n$ matrix representing the spacing of a set of n points in Euclidean space. If A is a Euclidean distance matrix and the points are defined on m-dimensional space, then the elements of A are given by

$$A = (a_{ij});$$
$$a_{ij} = ||x_i - x_j||_2^2$$

where $||.||_2$ denotes the 2-norm on R^m.

Properties

Simply put, the element a_{ij} describes the square of the distance between the i th and j th points in the set. By the properties of the 2-norm (or indeed, Euclidean distance in general), the matrix A has the following properties.

- All elements on the diagonal of A are zero (i.e. is it a hollow matrix).
- The trace of A is zero (by the above property).
- A is symmetric (i.e. $a_{ij} = a_{ji}$).
- $a_{ij}^{1/2} \leq a_{ik}^{1/2} + a_{kj}^{1/2}$ (by the triangle inequality)
- $a_{ij} \geq 0$
- The number of unique (distinct) non-zero values within an N-by-N Euclidean distance matrix is bounded (above) by $[N*(N-1)] / 2$ due to the matrix being symmetric and hollow.

Exchange Matrix

In mathematics, especially linear algebra, the exchange matrix is a special case of a permutation matrix, where the 1 elements reside on the counterdiagonal and all other elements are zero. In other words, it is a 'row-reversed' or 'column-reversed' version of the identity matrix.

$$J_2 = \begin{pmatrix} 0 & 1 \\ 1 & 0 \end{pmatrix}; \quad J_3 = \begin{pmatrix} 0 & 0 & 1 \\ 0 & 1 & 0 \\ 1 & 0 & 0 \end{pmatrix}; \quad J_n = \begin{pmatrix} 0 & 0 & \cdots & 0 & 0 & 1 \\ 0 & 0 & \cdots & 0 & 1 & 0 \\ 0 & 0 & \cdots & 1 & 0 & 0 \\ \vdots & \vdots & & \vdots & \vdots & \vdots \\ 0 & 1 & \cdots & 0 & 0 & 0 \\ 1 & 0 & \cdots & 0 & 0 & 0 \end{pmatrix}.$$

Definition

If J is an $n \times n$ exchange matrix, then the elements of J are defined such that:

$$J_{i,j} = \begin{cases} 1, & j = n - i + 1 \\ 0, & j \neq n - i + 1 \end{cases}$$

Properties

- $J^T = J$.
- $J^n = I$ for even n; $J^n = J$ for odd n, where n is any integer. Thus J is an involutary matrix; that is, $J^{-1} = J$.
- The trace of J is *1* if n is odd, and *0* if n is even.

Relationships

- Any matrix A satisfying the condition $AJ = JA$ is said to be centrosymmetric.
- Any matrix A satisfying the condition $AJ = JA^T$ is said to be persymmetric.

Fock Matrix

In the Hartree-Fock method of quantum mechanics, the Fock matrix is a matrix approximating the single-electron energy operator of a given quantum system in a given set of basis vectors [1].

It is most often formed in computational chemistry when attempting to solve the Roothaan equations for an atomic or molecular system. The Fock matrix is actually an approximation to the true Hamiltonian operator of the quantum system. It includes the effects of electron-electron repulsion only in an average way.

Importantly, because the Fock operator is a one-electron operator, it does not include the electron correlation energy.

The Fock matrix is defined by the *Fock operator*. For the restricted case which assumes closed-shell orbitals and single-determinantal wavefunctions, the Fock operator for the i-th electron is given by:

$$\hat{F}(i) = \hat{h}(i) + \sum_{j=1}^{n} [2\hat{J}_j(i) - \hat{K}_J(i)]$$

where:

$\hat{F}(i)$ is the Fock operator for the i-th electron in the system,

$\hat{h}(i)$ is the one-electron hamiltonian for the i-th electron,

n is the total number of occupied orbitals in the system (equal to $\left\lfloor \frac{N}{2} \right\rfloor$, where N is the number of electrons),

$\hat{J}_j(i)$ is the Coulomb operator, defining the repulsive force between the j-th and i-th electrons in the system,

$\hat{K}_j(i)$ is the exchange operator, defining the quantistic effect produced by exchanging two electrons. For systems with unpaired electrons there are many choices of Fock matrices.

Frobenius Matrix

A Frobenius matrix is a special kind of square matrix from numerical mathematics.

A matrix is a Frobenius matrix if it has the following three properties:

- all entries on the main diagonal are ones
- the entries below the main diagonal of at most one column are arbitrary
- every other entry is zero

The following matrix is an example.

$$A = \begin{pmatrix} 1 & 0 & 0 & \cdots & 0 \\ 0 & 1 & 0 & \cdots & 0 \\ 0 & a_{32} & 1 & \cdots & 0 \\ \vdots & \vdots & \vdots & \ddots & \vdots \\ 0 & a_{n2} & 0 & \cdots & 1 \end{pmatrix}$$

Frobenius matrices are invertible. The inverse of a Frobenius matrix is again a Frobenius matrix, equal to the original matrix with changed signs outside the main diagonal. The inverse of the example above is therefore:

$$A^{-1} = \begin{pmatrix} 1 & 0 & 0 & \cdots & 0 \\ 0 & 1 & 0 & \cdots & 0 \\ 0 & -a_{32} & 1 & \cdots & 0 \\ \vdots & \vdots & \vdots & \ddots & \vdots \\ 0 & -a_{n2} & 0 & \cdots & 1 \end{pmatrix}$$

Frobenius matrices are named after Ferdinand Georg Frobenius. An alternative name for this class of matrices is Gauss transformation, after Carl Friedrich Gauss. They are used in the process of Gaussian elimination to represent the Gaussian transformations.

If a matrix is multiplied from the left (left multiplied) with a Frobenius matrix, a linear combination of the remaining rows is added to a particular row of the matrix. Multiplication with the inverse matrix subtracts the corresponding linear combination from the given row.

This corresponds to one of the elementary operations of Gaussian elimination (besides the operation of transposing the rows and multiplying a row with a scalar multiple).

Fuzzy Associative Matrix

A fuzzy associative matrix expresses fuzzy logic rules in tabular form. These rules usually take two variables as input, mapping cleanly to a two-dimensional matrix, although theoretically a matrix of any number of dimensions is possible. Let's suppose we want to write fuzzy logic rules for a video game monster. We decide to start with two variables: hit points (HP) and firepower (FP). We might start with this:

HP/FP	*Very low HP*	*Low HP*	*Medium HP*	*High HP*	*Very high HP*
Very weak FP	Retreat!	Retreat!	Defend	Defend	Attack
Weak FP	Retreat!	Defend	Defend	Attack	Attack
Medium FP	Retreat!	Defend	Attack	Attack	Full attack!
High FP	Retreat!	Defend	Attack	Attack	Full attack!
Very high FP	Defend	Attack	Attack	Full attack!	Full attack!

This translates to:

```
    IF MonsterHP IS VeryLowHP AND MonsterFP IS VeryWeakFP
THEN Retreat

    IF MonsterHP IS LowHP AND MonsterFP IS VeryWeakFP
THEN Retreat

    IF MonsterHP IS MediumHP AND MonsterFP is VeryWeakFP
THEN Defend
```

...and so on. Multiple rules can fire at once, and very often, they will. The distinction between "very low" and "low" is probably fuzzy at many points, and at such points, both rules will fire. If it is more "very low" than it is low, then the "very low" rule will generate a stronger response. The program will evaluate all the rules that fire and use an appropriate defuzzification method to generate its actual response.

An implementation of this system might use either the matrix or the explicit IF/THEN form. The matrix makes it easy to visualize the system, but it also makes it impossible to add a third variable just for one rule, so it is less flexible.

There is no inherent pattern in the matrix. It appears as if the rules were just made up, and indeed they were. This is both a strength and a weakness of fuzzy logic in general.

It is often impractical or impossible to find an exact set of rules or formulae for dealing with a specific situation. For a sufficiently complex game, a mathematician would not be able to study the system and figure out a mathematically accurate set of rules.

However, this weakness is intrinsic to the realities of the situation, not of fuzzy logic itself. The strength of the system is that even if one of the rules is wrong, even greatly wrong, other rules that are correct are likely to fire as well and they may compensate for the error. This does not mean a fuzzy system should be sloppy. Depending on the system, it might get away with being sloppy, but it will underperform. While the rules are fairly arbitrary, they should be chosen carefully. If possible, an expert should decide on the rules, and the sets and rules should be tested vigorously and refined as needed. In this way, a fuzzy system is like an expert system. (Fuzzy logic is used in many true expert systems, as well.)

Gamma Matrices

In mathematical physics, the gamma matrices, $\{\gamma^0, \gamma^1, \gamma^2, \gamma^3\}$, also known as the Dirac matrices, are a set of conventional matrices with specific anticommutation relations that ensure they generate a matrix representation of the Clifford algebra $C\ell_{1,3}(\mathrm{R})$. It is also possible to define higher-dimensional gamma matrices.

When interpreted as the matrices of the action of a set of orthogonal basis vectors for contravariant vectors in Minkowski space, the column vectors on which the matrices act become a space of spinors, on which the Clifford algebra of spacetime acts. This in turn makes it possible to represent infinitesimal spatial rotations and Lorentz boosts. Spinors facilitate spacetime computations in general, and in particular are fundamental to the Dirac equation for relativistic spin-½ particles.

In Dirac representation, the four contravariant gamma matrices are

$$\gamma^0 = \begin{pmatrix} 1 & 0 & 0 & 0 \\ 0 & 1 & 0 & 0 \\ 0 & 0 & -1 & 0 \\ 0 & 0 & 0 & -1 \end{pmatrix}, \quad \gamma^1 = \begin{pmatrix} 0 & 0 & 0 & 1 \\ 0 & 0 & 1 & 0 \\ 0 & -1 & 0 & 0 \\ -1 & 0 & 0 & 0 \end{pmatrix}$$

$$\gamma^2 = \begin{pmatrix} 0 & 0 & 0 & -i \\ 0 & 0 & i & 0 \\ 0 & i & 0 & 0 \\ -i & 0 & 0 & 0 \end{pmatrix}, \quad \gamma^3 = \begin{pmatrix} 0 & 0 & 1 & 0 \\ 0 & 0 & 0 & -1 \\ -1 & 0 & 0 & 0 \\ 0 & 1 & 0 & 0 \end{pmatrix}.$$

Analogue sets of gamma matrices can be defined in any dimension and signature of the metric. For example the Pauli matrices are a set of "gamma" matrices in dimension 3 with metric of Euclidean signature (3,0).

Gell-Mann Matrices

The Gell-Mann matrices, named for Murray Gell-Mann, are one possible representation of the infinitesimal generators of the special unitary group called SU(3).

This group (a real Lie algebra in fact) has dimension eight and therefore it has some set with eight linearly independent generators, which can be written as g_i, with i taking values from 1 to 8. They obey the commutation relations

$$[g_i, g_j] = if^{ijk} g_k$$

where a sum over the index k is implied. The structure constants f^{ijk} are completely antisymmetric in the three indices and have values

$$f^{123} = 1, \quad f^{147} = f^{165} = f^{246} = f^{257} =$$

$$f^{345} = f^{376} = \frac{1}{2}, \quad f^{458} = f^{678} = \frac{\sqrt{3}}{2}.$$

Any set of Hermitian matrices which obey these relations are allowed. A particular choice of matrices is called a group representation, because any element of SU(3) can be written in the form $\exp(i\theta_j g_j)$, where θ_j are real numbers and a sum over the index j is implied.

Given one representation, another may be obtained by an arbitrary unitary transformation, since that leaves the commutator unchanged.

An important representation involves 3×3 matrices, because the group elements then act on complex vectors with 3 entries, i.e., on the fundamental representation of the group. A particular choice of this representation is

$$\lambda_1 = \begin{pmatrix} 0 & 1 & 0 \\ 1 & 0 & 0 \\ 0 & 0 & 0 \end{pmatrix} \quad \lambda_2 = \begin{pmatrix} 0 & -i & 0 \\ i & 0 & 0 \\ 0 & 0 & 0 \end{pmatrix} \quad \lambda_3 = \begin{pmatrix} 1 & 0 & 0 \\ 0 & -1 & 0 \\ 0 & 0 & 0 \end{pmatrix}$$

$$\lambda_4 = \begin{pmatrix} 0 & 0 & 1 \\ 0 & 0 & 0 \\ 1 & 0 & 0 \end{pmatrix} \quad \lambda_5 = \begin{pmatrix} 0 & 0 & -i \\ 0 & 0 & 0 \\ i & 0 & 0 \end{pmatrix}$$

$$\lambda_6 = \begin{pmatrix} 0 & 0 & 0 \\ 0 & 0 & 1 \\ 0 & 1 & 0 \end{pmatrix} \lambda_7 = \begin{pmatrix} 0 & 0 & 0 \\ 0 & 0 & -i \\ 0 & i & 0 \end{pmatrix} \lambda_8 = \frac{1}{\sqrt{3}} \begin{pmatrix} 1 & 0 & 0 \\ 0 & 1 & 0 \\ 0 & 0 & -2 \end{pmatrix}$$

and $g_i = \lambda_i / 2$. These matrices are traceless, Hermitian, and obey the extra relation $\mathrm{tr}(\lambda_i \lambda_j) = 2\delta_{ij}$ These properties were chosen by Gell-Mann because they then generalize the Pauli matrices. In this representation it is clear that the Cartan subalgebra is the set of linear combinations (with real coefficients) of the two matrices λ_3 and λ_8, which commute $\{\lambda_1, \lambda_2, x\}$, $\{\lambda_4, \lambda_5, y\}$, and $\{\lambda_6, \lambda_7, z\}$, where the *x*, *y*, *z* must consist of linear combinations of λ_3 and λ_8.

These matrices form a useful representation for computations in the quark model, and, to a lesser extent, in quantum chromodynamics.

Generalized Inverse

In mathematics, a generalized inverse or pseudoinverse of a matrix *A* is a matrix that has some properties of the inverse matrix of *A* but not necessarily all of them. The term "the pseudoinverse" commonly means the Moore–Penrose pseudoinverse.

The purpose of constructing a generalized inverse is to obtain a matrix that can serve as the inverse in some sense for a wider class of matrices than invertible ones. Typically, the generalized inverse exists for an arbitrary matrix, and when a matrix has an inverse, then its inverse and the generalized inverse are the same. Some generalized inverses can be defined in any mathematical structure that involves associative multiplication, that is, in a semigroup.

Givens Rotation

In numerical linear algebra, a Givens rotation is a rotation in the plane spanned by two coordinates axes. Givens rotations are named after Wallace Givens, who introduced them to numerical analysts in the 1950s while he was working at Argonne National Laboratory.

Matrix Representation

A Givens rotation is represented by a matrix of the form

$$G(i,j,\theta)=\begin{bmatrix} 1 & \cdots & 0 & \cdots & 0 & \cdots & 0 \\ \vdots & \ddots & \vdots & & \vdots & & \vdots \\ 0 & \cdots & c & \cdots & -s & \cdots & 0 \\ \vdots & & \vdots & \ddots & \vdots & & \vdots \\ 0 & \cdots & s & \cdots & c & \cdots & 0 \\ \vdots & & \vdots & & \vdots & \ddots & \vdots \\ 0 & \cdots & 0 & \cdots & 0 & \cdots & 1 \end{bmatrix}$$

where $c = \cos(\theta)$ and $s = \sin(\theta)$ appear at the intersections ith and jth rows and columns. That is, the non-zero elements of Givens matrix is given by:

$$\begin{aligned} g_{kk} &= 1 \qquad \text{for } k \neq i,j \\ g_{ii} &= c \\ g_{jj} &= c \\ g_{ji} &= -s \\ g_{ij} &= s \qquad \text{for } i > j \end{aligned}$$

(sign of sine switches for j > i)

The product $G(i,j,\theta)$x represents a counterclockwise rotation of the vector x in the (i,j) plane of θ radians, hence the name Givens rotation. The main use of Givens rotations in numerical linear algebra is to introduce zeros in vectors or matrices. This effect can, for example, be employed for computing the QR decomposition of a matrix. One advantage over Householder transformations is that they can easily be parallelised, and another is that often for very sparse matrices they have a lower operation count.

Stable Calculation

When a Givens rotation matrix, $G(i,k,\theta)$, multiplies another matrix, A, from the left, GA, only rows i and j of A are affected. Thus we restrict attention to the following problem. Given a and b, find $c = \cos\theta$ and $s = \sin\theta$ such that

$$\begin{bmatrix} c & -s \\ s & c \end{bmatrix}\begin{bmatrix} a \\ b \end{bmatrix} = \begin{bmatrix} r \\ 0 \end{bmatrix}.$$

Explicit calculation of θ is rarely necessary or desirable. Instead we directly seek *c*, *s*, and *r*. An obvious solution would be

$$r \leftarrow \sqrt{a^2 + b^2}$$
$$c \leftarrow a / r$$
$$s \leftarrow -b / r.$$

However, the computation for *r* may overflow or underflow. An alternative formulation avoiding this problem (Golub & Van Loan 1996, §5.1.8) is implemented as the hypot function in many programming languages.

Furthermore, as Anderson (2000) discovered in improving LAPACK, a previously overlooked numerical consideration is continuity. To achieve this, we require *r* to be positive.

```
if (b=0) then {c ← copysign(1,a); s ¬ 0; r ← abs(a)}
 else if (a = 0) then {c ← 0; s ← copysign(1,b); r ←
abs(b)}
 else if (abs(b) > abs(a)) then
   t ← a/b
   u ← copysign(sqrt(1+t*t),b)
   s ← 1/u
   c ← s*t
   r ← b*u
else
   t ← b/a
   u ← copysign(sqrt(1+t*t),a)
   c ← 1/u
   s ← c*t
   r ← a*u
```

This is written in terms of the IEEE 754 copysign(x,y) function, which provides a safe and cheap way to copy the sign of y to x. If that is not available, x*sgn(y), using the sign function, is an alternative.

Triangularization

Given the following 3x3 Matrix, perform two iterations of the Given's Rotation to bring the matrix to an upper Triangular matrix in order to compute the QR decomposition.

$$A = \begin{bmatrix} 6 & 5 & 0 \\ 5 & 1 & 4 \\ 0 & 4 & 3 \end{bmatrix}$$

In order to form the desired matrix, we must zero elements (2,1) and (3,2). We first select element (2,1) to zero. Using a rotation matrix of:

$$G_1 = \begin{bmatrix} c & -s & 0 \\ s & c & 0 \\ 0 & 0 & 1 \end{bmatrix}$$

We have the following matrix multiplication:

$$\begin{bmatrix} c & -s & 0 \\ s & c & 0 \\ 0 & 0 & 1 \end{bmatrix} \begin{bmatrix} 6 & 5 & 0 \\ 5 & 1 & 4 \\ 0 & 4 & 3 \end{bmatrix}$$

Where:

$$\begin{aligned} r &= \sqrt{6^2 + 5^2} = 7.8102 \\ c &= 6 / r = 0.7682 \\ s &= -5 / r = -0.6402 \end{aligned}$$

Plugging in these values for c and s and performing the matrix multiplication above yields a new A of:

$$A = \begin{bmatrix} 7.8102 & 4.4813 & 2.5607 \\ 0 & -2.4327 & 3.0729 \\ 0 & 4 & 3 \end{bmatrix}$$

We now want to zero element (3,2) to finish off the process. Using the same idea as before, we have a rotation matrix of:

$$G_2 = \begin{bmatrix} 1 & 0 & 0 \\ 0 & c & -s \\ 0 & s & c \end{bmatrix}$$

We are presented with the following matrix multiplication:

$$\begin{bmatrix} 1 & 0 & 0 \\ 0 & c & -s \\ 0 & s & c \end{bmatrix} \begin{bmatrix} 7.8102 & 4.4813 & 2.5607 \\ 0 & -2.4327 & 3.0729 \\ 0 & 4 & 3 \end{bmatrix}$$

Where:

$$\begin{aligned} r &= \sqrt{(-2.4327)^2 + 4^2} = 4.6817 \\ c &= -2.4327 / r = -0.5196 \\ s &= -4 / r = -0.8544 \end{aligned}$$

Plugging in these values for c and s and performing the multiplications gives us a new matrix of:

$$R = \begin{bmatrix} 7.8102 & 4.4813 & 2.5607 \\ 0 & 4.6817 & 0.9664 \\ 0 & 0 & -4.1843 \end{bmatrix}$$

This new matrix R is the upper triangular matrix needed to perform an iteration of the QR decomposition.

Q is now formed using the transpose of the rotation matrices in the following manner:

$$Q = G_1^T * G_2^T$$

Performing this matrix multiplication yields:

$$Q = \begin{bmatrix} 0.7682 & 0.3327 & 0.5470 \\ 0.6402 & -0.3992 & -0.6564 \\ 0 & 0.8544 & -0.5196 \end{bmatrix}$$

This completes two iterations of the Givens Rotation and calculating the QR decomposition can now be done.

Givens Rotations in Clifford Algebras

In Clifford algebras and its child structures like geometric algebra rotations are represented by bivectors.

Givens rotations are represented by the external product of the basis vectors. Given any pair of basis vectors e_i, e_j Givens rotations bivectors are:

$$B_{ij} = e_i \wedge e_j$$

Their action on any vector is written :

$$v = e^{-(\theta/2)(e_i \wedge e_j)} u e^{(\theta/2)(e_i \wedge e_j)}$$

where :

$$e^{(\theta/2)(e_i \wedge e_j)} = \cos(\theta / 2) + \sin(\theta / 2) e_i \wedge e_j$$

Dimension 3

There are three Givens rotations in dimension 3:

$$R_X(\theta) = \begin{bmatrix} 1 & 0 & 0 \\ 0 & \cos\theta & -\sin\theta \\ 0 & \sin\theta & \cos\theta \end{bmatrix}$$

$$R_Y(\theta) = \begin{bmatrix} \cos\theta & 0 & \sin\theta \\ 0 & 1 & 0 \\ -\sin\theta & 0 & \cos\theta \end{bmatrix}$$

$$R_Z(\theta) = \begin{bmatrix} \cos\theta & -\sin\theta & 0 \\ \sin\theta & \cos\theta & 0 \\ 0 & 0 & 1 \end{bmatrix}$$

Given that they are endomorphisms they can be composed with each other as many times as desired, keeping in mind that $g \neq f$ `– $f \neq g$. These three Givens rotations composed can generate any rotation matrix.

This means that they can transform the basis of the space to any other frame in the space. When rotations are performed in the right order, the values of the rotation angles of the final frame will be equal to the Euler angles of the final frame in the corresponding convention.

For example, an operator $R = R_Y(\theta_3).R_X(\theta_2).R_Z(\theta_1)$ transforms the basis of the space into a frame with angles roll, pitch and yaw $YPR = (\theta_3, \theta_2, \theta_1)$ in the Tait-Bryan convention z-x-y (convention in which the line of nodes is perpendicular to z and Y axes, also named Y-X'-Z").

For the same reason, any rotation matrix in 3D can be decomposed in a product of three of these rotations.

The meaning of the composition of two Givens rotations g≠f is an operator that transforms vectors first by f and then by g, being f and g rotations about one axis of basis of the space. This is similar to the extrinsic rotation equivalence for Euler angles.

Gramian Matrix

In linear algebra, the Gramian matrix (or Gram matrix or Gramian) of a set of vectors $v_1,\ldots,v_n$ in an inner product space is the Hermitian matrix of inner products, whose entries are given by $G_{ij} = \langle v_j, v_i \rangle$. An important application is to compute linear independence: a set of vectors is linearly independent if and only if the Gram determinant (the determinant of the Gram matrix) is non-zero. It is named after Jørgen Pedersen Gram.

Examples

Most commonly, the vectors are elements of an Euclidean space, or are functions in an L^2 space, such as continuous functions on a compact interval $[a, b]$ (which are a subspace of $L^2([a, b])$). Given real-valued functions $\{\ell_i(\cdot), i = 1,\ldots,n\}$ on the interval $[t_0, t_f]$, the Gram matrix $G = [G_{ij}]$, is given by the standard inner product on functions:

$$G_{ij} = \int_{t_0}^{t_f} \ell_i(\tau)\overline{\ell_j}(\tau)d\tau.$$

Given a real matrix A, the matrix A^TA is a Gram matrix (of the columns of A), while the matrix AA^T is the Gram matrix of the rows of A. For a general bilinear form B on a finite-dimensional vector space over any field we can define a Gram matrix G attached to a set of vectors $v_1,\ldots,v_n$ by $G_{ij} = B(v_i, v_j)$. The matrix will be symmetric if the bilinear form B is symmetric.

Applications

- If the vectors are centred random variables, the Gramian is proportional to the covariance matrix, with the scaling determined by the number of elements in the vector.
- In quantum chemistry, the Gram matrix of a set of basis vectors is the overlap matrix.
- In control theory (or more generally systems theory), the controllability Gramian and observability Gramian determine properties of a linear system.
- Gramian matrices arise in covariance structure model fitting.

- In the finite element method, the Gram matrix arises from approximating a function from a finite dimensional space; the Gram matrix entries are then the inner products of the basis functions of the finite dimensional subspace.
- In machine learning, kernel functions are often represented as Gram matrices.

Properties

Positive Semidefinite

The Gramian matrix is positive semidefinite, and every positive semidefinite matrix is the Gramian matrix for some set of vectors. Further, in finite-dimensions it determines the vectors up to isomorphism, i.e. any two sets of vectors with the same Gramian matrix must be related by a single unitary matrix. The Gramian matrix of any orthonormal basis is the identity matrix. The infinite-dimensional analog of this statement is Mercer's theorem.

Change of Basis

Under change of basis represented by an invertible matrix P, the Gram matrix will change by a matrix congruence to $P^{T}GP$.

Gram Determinant

The Gram determinant or Gramian is the determinant of the Gram matrix:

$$G(x_1,\ldots,x_n) = \begin{vmatrix} \langle x_1,x_1\rangle & \langle x_1,x_2\rangle & \cdots & \langle x_1,x_n\rangle \\ \langle x_2,x_1\rangle & \langle x_2,x_2\rangle & \cdots & \langle x_2,x_n\rangle \\ \vdots & \vdots & \ddots & \vdots \\ \langle x_n,x_1\rangle & \langle x_n,x_2\rangle & \cdots & \langle x_n,x_n\rangle \end{vmatrix}.$$

Geometrically, the Gram determinant is the square of the volume of the parallelotope formed by the vectors. In particular, the vectors are linearly independent if and only if the Gram determinant is nonzero (if and only if the Gram matrix is nonsingular).

The Gram determinant can also be expressed in terms of the exterior product of vectors by

$$G(x_1,\ldots,x_n) = \| x_1 \wedge \cdots \wedge x_n \|^2 .$$

Chapter 9

Hadamard Matrix

In mathematics, an Hadamard matrix, named after the French mathematician Jacques Hadamard, is a square matrix whose entries are either +1 or −1 and whose rows are mutually orthogonal. In geometric terms, this means that every two different rows in a Hadamard matrix represent two perpendicular vectors, while in combinatorial terms, it means that every two different rows have matching entries in exactly half of their columns and mismatched entries in the remaining columns.

It is a consequence of this definition that the corresponding properties hold for columns as well as rows. The n-dimensional parallelotope spanned by the rows of an $n\times n$ Hadamard matrix has the maximum possible n-dimensional volume among parallelotopes spanned by vectors whose entries are bounded in absolute value by 1.

Equivalently, a Hadamard matrix has maximal determinant among matrices with entries of absolute value less than or equal to 1 and so, is an extremal solution of Hadamard's maximal determinant problem. Certain Hadamard matrices can almost directly be used as an error-correcting code using a Hadamard code (generalized in Reed–Muller codes), and are also used in balanced repeated replication (BRR), used by statisticians to estimate the variance of a parameter estimator.

Properties

It follows from the definition that a Hadamard matrix H of order n satisfies

$$HH^{\mathrm{T}} = nI_n$$

where I_n is the $n \times n$ identity matrix and H^T is the transpose of H. Consequently the determinant of H equals $\pm n^{n/2}$. Suppose that M is a complex matrix of order n, whose entries are bounded by $|M_{ij}| \leq 1$, for each i, j between 1 and n. Then Hadamard's determinant bound states that

$$|\det(M)| \leq n^{n/2}.$$

Equality in this bound is attained for a real matrix M if and only if M is a Hadamard matrix. The order of a Hadamard matrix must be 1, 2, or a multiple of 4.

Sylvester's Construction

Examples of Hadamard matrices were actually first constructed by James Joseph Sylvester in 1867. Let H be a Hadamard matrix of order n. Then the partitioned matrix

$$\begin{bmatrix} H & H \\ H & -H \end{bmatrix}$$

is a Hadamard matrix of order $2n$. This observation can be applied repeatedly and leads to the following sequence of matrices, also called Walsh matrices.

$$H_1 = \begin{bmatrix} 1 \end{bmatrix},$$

$$H_2 = \begin{bmatrix} 1 & 1 \\ 1 & -1 \end{bmatrix},$$

and

$$H_{2^k} = \begin{bmatrix} H_{2^{k-1}} & H_{2^{k-1}} \\ H_{2^{k-1}} & -H_{2^{k-1}} \end{bmatrix} = H_2 \otimes H_{2^{k-1}},$$

for $2 \leq k \in N$, where $\otimes$ denotes the Kronecker product.

In this manner, Sylvester constructed Hadamard matrices of order 2^k for every non-negative integer k.

Sylvester's matrices have a number of special properties. They are symmetric and have trace zero. The elements in the first column and the first row are all positive. The elements in all the other rows and columns are evenly divided between positive and negative. Sylvester matrices are closely connected with Walsh functions.

Alternative Construction

If we map the elements of the Hadamard matrix using the group homomorphism $\{1,-1,\times\} \mapsto \{0,1,\oplus\}$, we can describe an alternative construction of Sylvester's Hadamard matrix. First consider the matrix F_n, the $n \times 2^n$ matrix whose columns consist of all n-bit numbers arranged in ascending counting order. We may define F_n recursively by

$$F_1 = \begin{bmatrix} 0 & 1 \end{bmatrix}$$

$$F_n = \begin{bmatrix} 0_{1\times 2^{n-1}} & 1_{1\times 2^{n-1}} \\ F_{n-1} & F_{n-1} \end{bmatrix}.$$

It can be shown by induction that the image of the Hadamard matrix under the above homomorphism is given by

$$H_{2^n} = F_n^{\mathrm{T}} F_n.$$

This construction demonstrates that the rows of the Hadamard matrix H_{2^n} can be viewed as a length 2^n linear error-correcting code of rank n, and minimum distance 2^{n-1} with generating matrix F_n. This code is also referred to as a Walsh code. The Hadamard code, by contrast, is constructed from the Hadamard matrix by a slightly different procedure.

The Hadamard Conjecture

The most important open question in the theory of Hadamard matrices is that of existence. The Hadamard conjecture proposes that a Hadamard matrix of order $4k$ exists for every positive integer k.

A generalization of Sylvester's construction proves that if H_n and H_m are Hadamard matrices of orders n and m respectively, then $H_n \otimes H_m$ is an Hadamard matrix of order nm. This result is used to produce Hadamard matrices of higher order once those of smaller orders are known. Sylvester's 1867 construction yields Hadamard matrices of order 1, 2, 4, 8, 16, 32, etc. Hadamard matrices of orders 12 and 20 were subsequently constructed by Hadamard (in 1893). In 1933 Raymond Paley discovered a construction that produces a

Hadamard matrix of order $q+1$ when q is any prime power that is congruent to 3 modulo 4 and that produces a Hadamard matrix of order $2(q+1)$ when q is a prime power that is congruent to 1 modulo 4. His method uses finite fields. The Hadamard conjecture should probably be attributed to Paley. The smallest order that cannot be constructed by a combination of Sylvester's and Paley's methods is 92. A Hadamard matrix of this order was found using a computer by Baumert, Golomb, and Hall in 1962. They used a construction, due to Williamson, that has yielded many additional orders.

Many other methods for constructing Hadamard matrices are now known. In 2005, Hadi Kharaghani and Behruz Tayfeh-Rezaie published their construction of a Hadamard matrix of order 428. As a result, the smallest order for which no Hadamard matrix is presently known is 668. As of 2008, there are 13 multiples of 4 less than or equal to 2000 for which no Hadamard matrix of that order is known. They are: 668, 716, 892, 1004, 1132, 1244, 1388, 1436, 1676, 1772, 1916, 1948, 1964.

Equivalence of Hadamard Matrices

Two Hadamard matrices are considered equivalent if one can be obtained from the other by negating rows or columns, or by interchanging rows or columns. Up to equivalence, there is a unique Hadamard matrix of orders 1, 2, 4, 8, and 12. There are 5 inequivalent matrices of order 16, 3 of order 20, 60 of order 24, and 487 of order 28. Millions of inequivalent matrices are known for orders 32, 36, and 40. Using a coarser notion of equivalence that also allows transposition, there are 4 inequivalent matrices of order 16, 3 of order 20, 36 of order 24, and 294 of order 28 .

Skew Hadamard Matrices

A Hadamard matrix H is *skew* if $H^{\mathrm{T}} + H = 2I$.

Reid and Brown in 1972 showed that there exists a "doubly regular tournament of order n– if and only if there exists a skew Hadamard matrix of order n+1.

Generalizations and Special Cases

Many generalizations and special cases of Hadamard matrices have been investigated in the mathematical literature.

One basic generalization is the weighing matrix, a square matrix in which entries may also be zero and which satisfies $WW^T = wI$ for some w, its weight. A weighing matrix with its weight equal to its order is a Hadamard matrix.

Another generalization defines a complex Hadamard matrix to be a matrix in which the entries are complex numbers of unit modulus and which satisfies $H H^* = n I_n$ where H^* is the conjugate transpose of *H*. Complex Hadamard matrices arise in the study of operator algebras and the theory of quantum computation. Butson-type Hadamard matrices are complex Hadamard matrices in which the entries are taken to be q^{th} roots of unity. The term "complex Hadamard matrix– has been used by some authors to refer specifically to the case $q = 4$. Regular Hadamard matrices are real Hadamard matrices whose row and column sums are all equal. A necessary condition on the existence of a regular $n \times n$ Hadamard matrix is that *n* be a perfect square. A circulant matrix is manifestly regular, and therefore a circulant Hadamard matrix would have to be of perfect square order. Moreover, if an $n \times n$ circulant Hadamard matrix existed with $n>1$ then *n* would necessarily have to be of the form $4u^2$ with *u* odd.

The circulant Hadamard matrix conjecture, however, asserts that, apart from the known 1×1 and 4×4 examples, no such matrices exist. This was verified for all but 26 values of *u* less than 10^4.

Practical Applications

- *Olivia MFSK:* an amateur-radio digital protocol designed to work in difficult (low signal-to-noise ratio plus multipath propagation) conditions on shortwave bands.
- *Balanced Repeated Replication (BRR):* a technique used by statisticians to estimate the variance of a statistical estimator.
- *Coded aperture spectrometry:* an instrument for measuring the spectrum of light. The mask element used in coded aperture spectrometers is often a variant of a Hadamard matrix.
- *Feedback Delay Networks:* Digital reverberation devices which use Hadamard matrices to blend sample values

- *Plackett:* Burman design of experiments for investigating the dependence of some measured quantity on a number of independent variables.

Hadamard's Maximal Determinant Problem

Hadamard's maximal determinant problem, named after Jacques Hadamard, asks for the largest determinant of a matrix with elements equal to 1 or −1. The analogous question for matrices with elements equal to 0 or 1 is equivalent since the maximal determinant of a {1,–1} matrix of size n is 2^{n-1} times the maximal determinant of a {0,1} matrix of size n–1.

The problem was posed by Hadamard in the 1893 paper in which he presented his famous determinant bound and remains unsolved for matrices of general size. Hadamard's bound implies that {1, −1}-matrices of size n have determinant at most $n^{n/2}$. Hadamard observed that a construction of Sylvester produces examples of matrices that attain the bound when n is a power of 2, and produced examples of his own of sizes 12 and 20. He also showed that the bound is not attainable when n is not equal to 1, 2, or a multiple of 4.

Additional examples were later constructed by Scarpis and Paley and subsequently by many other authors. Such matrices are now known as Hadamard matrices. They have received intensive study.

Matrix sizes n for which n a– 1, 2, or 3 (mod 4) have received less attention. The earliest results are due to Barba, who tightened Hadamard's bound for n odd, and Williamson, who found the largest determinants for n=3, 5, 6, and 7. Some important results include

- Tighter bounds, due to Barba, Ehlich, and Wojtas, for n a– 1, 2, or 3 (mod 4), which, however, are known not to be always attainable,
- A few infinite sequences of matrices attaining the bounds for n a– 1 or 2 (mod 4),
- A number of matrices attaining the bounds for specific n a–1 or 2 (mod 4),
- A number of matrices not attaining the bounds for specific n a– 1 or 3 (mod 4), but that have been proved by exhaustive computation to have maximal determinant.

The design of experiments in statistics makes use of {1, –1} matrices X (not necessarily square) for which the information matrix $X^T X$ has maximal determinant. (The notation X^T denotes the transpose of X.) Such matrices are known as D-optimal designs. If X is a square matrix, it is known as a saturated D-optimal design.

Hadamard Matrices

Any two rows of an $n \times n$ Hadamard matrix are orthogonal, which is impossible for a {1, –1} matrix when n is an odd number. When n a– 2 (mod 4), two rows that are both orthogonal to a third row cannot be orthogonal to each other. Together, these statements imply that an $n \times n$ Hadamard matrix can exist only if $n = 1$, 2, or a multiple of 4. Hadamard matrices have been well studied, but it is not known whether a Hadamard matrix of size $4k$ exists for every $k \geq 1$. The smallest n for which an $n \times n$ Hadamard matrix is not known to exist is 668.

Equivalence and Normalization of {1, –1} Matrices

Any of the following operations, when performed on a {1, –1} matrix R, changes the determinant of R only by a minus sign:

- Negation of a row.
- Negation of a column.
- Interchange of two rows.
- Interchange of two columns.

Two {1,–1} matrices, R_1 and R_2, are considered equivalent if R_1 can be converted to R_2 by some sequence of the above operations.

The determinants of equivalent matrices are equal, except possibly for a sign change, and it is often convenient to standardize R by means of negations and permutations of rows and columns.

A {1, –1} matrix is normalized if all elements in its first row and column equal 1.

When the size of a matrix is odd, it is sometimes useful to use a different normalization in which every row and column contains an even number of elements 1 and an odd number of elements –1. Either of these normalizations can be accomplished using the first two operations.

Connection of the Maximal Determinant Problems for {1, •1} and {0, 1} Matrices

There is a one-to-one map from the set of normalized $n \times n$ {1, •1} matrices to the set of $(n-1)\times(n-1)$ {0, 1} matrices under which the magnitude of the determinant is reduced by a factor of 2^{1-n}.

This map consists of the following steps.

1. Subtract row 1 of the {1, –1} matrix from rows 2 through n. (This does not change the determinant.)
2. Extract the $(n-1)\times(n-1)$ submatrix consisting of rows 2 through n and columns 2 through n.

   This matrix has elements 0 and –2. (The determinant of this submatrix is the same as that of the original matrix, as can be seen by performing a cofactor expansion on column 1 of the matrix obtained in Step 1.)
3. Divide the submatrix by –2 to obtain a {0, 1} matrix. (This multiplies the determinant by $(-2)^{1-n}$.)

Example:

$$\begin{bmatrix} 1 & 1 & 1 & 1 \\ 1 & -1 & -1 & 1 \\ 1 & 1 & -1 & -1 \\ 1 & -1 & 1 & -1 \end{bmatrix} \to \begin{bmatrix} 1 & 1 & 1 & 1 \\ 0 & -2 & -2 & 0 \\ 0 & 0 & -2 & -2 \\ 0 & -2 & 0 & -2 \end{bmatrix}$$

$$\to \begin{bmatrix} -2 & -2 & 0 \\ 0 & -2 & -2 \\ -2 & 0 & -2 \end{bmatrix} \to \begin{bmatrix} 1 & 1 & 0 \\ 0 & 1 & 1 \\ 1 & 0 & 1 \end{bmatrix}$$

In this example, the original matrix has determinant –16 and its image has determinant $2 = -16 \cdot (-2)^{-3}$. Since the determinant of a {0, 1} matrix is an integer, the determinant of an $n \times n$ {1, –1} matrix is an integer multiple of 2^{n-1}.

Upper Bounds on the Maximal Determinant

Gram Matrix

Let R be an n by n {1, –1} matrix. The Gram matrix of R is defined to be the matrix $G = RR^{\mathrm{T}}$. From this definition it follows that G

1. is an integer matrix,
2. is symmetric,
3. is positive-semidefinite,
4. has constant diagonal whose value equals n.

Negating rows of R or applying a permutation to them results in the same negations and permutation being applied both to the rows, and to the corresponding columns, of G. We may also define the matrix $G2 = R^T R$. The matrix G is the usual Gram matrix of a set of vectors, derived from the set of rows of R, while $G2$ is the Gram matrix derived from the set of columns of R. A matrix R for which $G = G2$ is a normal matrix. Every known maximal-determinant matrix is equivalent to a normal matrix, but it is not known whether this is always the case.

Hadamard's Bound (for all n)

Hadamard's bound can be derived by noting that $|\det R| = (\det G)^{1/2} \leq (\det nI)^{1/2} = n^{n/2}$, which is a consequence of the observation that nI, where I is the n by n identity matrix, is the unique matrix of maximal determinant among matrices satisfying properties 1–4. That $\det R$ must be an integer multiple of 2^{n-1} can be used to provide another demonstration that Hadamard's bound is not always attainable. When n is odd, the bound $n^{n/2}$ is either non-integer or odd, and is therefore unattainable except when $n = 1$. When $n = 2k$ with k odd, the highest power of 2 dividing Hadamard's bound is 2^k which is less than 2^{n-1} unless $n = 2$. Therefore Hadamard's bound is unattainable unless $n = 1$, 2, or a multiple of 4.

Barba's Bound for n Odd

When n is odd, property 1 for Gram matrices can be strengthened to

1. G is an odd-integer matrix.

This allows a sharper upper bound to be derived: $|\det R| = (\det G)^{1/2} \leq (\det (n\text{-}1)I{+}J)^{1/2} = (2n-1)^{1/2}(n-1)^{(n-1)/2}$, where J is the all-one matrix. Here $(n\text{-}1)I{+}J$ is the maximal-determinant matrix satisfying the modified property 1 and properties 2–4. It is unique up to multiplication of any set of rows and the corresponding set of columns by –1. The bound is not attainable unless $2n-1$ is a perfect square, and is therefore never attainable when n a– 3 (mod 4).

The Ehlich–Wojtas Bound for n a– 2 *(mod 4)*

When n is even, the set of rows of R can be partitioned into two subsets.

- Rows of even type contain an even number of elements 1 and an even number of elements -1.
- Rows of odd type contain an odd number of elements 1 and an odd number of elements -1.

The dot product of two rows of the same type is congruent to n (mod 4); the dot product of two rows of opposite type is congruent to $n+2$ (mod 4). When n a– 2 (mod 4), this implies that, by permuting rows of R, we may assume the standard form,

$$G = \begin{bmatrix} A & B \\ B^{T} & D \end{bmatrix},$$

where A and D are symmetric integer matrices whose elements are congruent to 2 (mod 4) and B is a matrix whose elements are congruent to 0 (mod 4). In 1964, Ehlich and Wojtas independently showed that in the maximal determinant matrix of this form, A and D are both of size $n/2$ and equal to $(n-2)I+2J$ while B is the zero matrix.

This optimal form is unique up to multiplication of any set of rows and the corresponding set of columns by -1 and to simultaneous application of a permutation to rows and columns. This implies the bound det R £ $(2n\text{-}2)(n\text{-}2)^{(n\text{-}2)/2}$.

Ehlich showed that if R attains the bound, and if the rows and columns of R are permuted so that both $G = RR^{T}$ and $G2 = R^{T}R$ have the standard form and are suitably normalized, then we may write

$$R = \begin{bmatrix} W & X \\ Y & Z \end{bmatrix}$$

where W, X, Y, and Z are $(n/2)\times(n/2)$ matrices with constant row and column sums w, x, y, and z that satisfy $z = -w$, $y = x$, and $w^2+x^2 = 2n-2$. Hence the Ehlich–Wojtas bound is not attainable unless $2n-2$ is expressible as the sum of two squares.

Ehlich's Bound for n a– 3 (mod 4)

When n is odd, then by using the freedom to multiply rows by -1, one may impose the condition that each row of R contain an even number of elements 1 and an odd number of elements -1. It can be shown that, if this normalization is assumed, then property 1 of G may be strengthened to *1.* G is a matrix with integer elements congruent to n (mod 4).

When n a– 1 (mod 4), the optimal form of Barba satisfies this stronger property, but when n a– 3 (mod 4), it does not. This means that the bound can be sharpened in the latter case. Ehlich showed that when n a– 3 (mod 4), the strengthened property 1 implies that the maximal-determinant form of G can be written as $B–J$ where J is the all-one matrix and B is a block-diagonal matrix whose diagonal blocks are of the form $(n\text{-}3)I+4J$. Moreover, he showed that in the optimal form, the number of blocks, s, depends on n as shown in the table below, and that each block either has size r or size $r+1$ where $r = \lfloor n/s \rfloor$.

n	s
3	3
7	5
11	5 or 6
15 – 59	6
≥ 63	7

Except for n=11 where there are two possibilities, the optimal form is unique up to multiplication of any set of rows and the corresponding set of columns by -1 and to simultaneous application of a permutation to rows and columns. This optimal form leads to the bound

$$\det R \le (n-3)^{(n-s)/2}(n-3+4r)^{u/2}(n+1+4r)^{v/2}$$
$$\left[1-\frac{ur}{n-3+4r}-\frac{v(r+1)}{n+1+4r}\right]^{1/2},$$

where $v = n–rs$ is the number of blocks of size r+1 and $u = s–v$ is the number of blocks of size r. Cohn analyzed the bound and determined that, apart from $n = 3$, it is an integer only for $n = 112t^2 \pm 28t+7$ for some positive integer t. Tamura derived additional restrictions on the attainability of the bound using the Hasse-Minkowski theorem on the rational equivalence of quadratic forms, and showed that the smallest $n > 3$ for which Ehlich's bound is conceivably attainable is 511.

Maximal Determinants up to Size 18

The maximal determinants of {1,–1} matrices up to size $n = 18$ are given in the following table. Size 19 is the smallest open case.

In the table, $D(n)$ represents the maximal determinant divided by 2^{n-1}. Equivalently, $D(n)$ represents the maximal determinant of a {0, 1} matrix of size $n-1$.

n	*D(n)*	*Notes*
1	1	Hadamard matrix
2	1	Hadamard matrix
3	1	Attains Ehlich bound
4	2	Hadamard matrix
5	3	Attains Barba bound; circulant matrix
6	5	Attains Ehlich–Wojtas bound
7	9	98.20% of Ehlich bound
8	32	Hadamard matrix
9	56	84.89% of Barba bound
10	144	Attains Ehlich–Wojtas bound
11	320	94.49% of Ehlich bound; three non-equivalent matrices
12	1458	Hadamard matrix
13	3645	Attains Barba bound; maximal-determinant matrix is {1,–1} incidence matrix of projective plane of order 3
14	9477	Attains Ehlich–Wojtas bound
15	25515	97.07% of Ehlich bound
16	131072	Hadamard matrix; five non-equivalent matrices
17	327680	87.04% of Barba bound; three non-equivalent matrices
18	1114112	Attains Ehlich–Wojtas bound; three non-equivalent matrices

Hat Matrix

In statistics, the hat matrix, *H*, sometimes also called projection matrix, maps the vector of observed values to the vector of fitted values. It describes the influence each observed value has on each fitted value. The diagonal elements of the hat matrix are the leverages, which describe the influence each observed value has on the fitted value for that same observation.

If the vector of observed values is denoted by y and the vector of fitted values by w,

$$\hat{\mathbf{y}} = H\mathbf{y}.$$

As w is usually pronounced "y-hat–, the hat matrix is so named as it "puts a hat on y".

Suppose that we wish to solve a linear model using linear least squares. The model can be written as

$$\mathbf{y} = X\boldsymbol{\beta} + \boldsymbol{\varepsilon},$$

where X is a matrix of explanatory variables (the design matrix), β is a vector of unknown parameters to be estimated, and ε is the error vector.

Uncorrelated Errors

For uncorrelated errors, the estimated parameters are

$$\hat{\boldsymbol{\beta}} = \left(X^\top X\right)^{-1} X^\top \mathbf{y},$$

so the fitted values are

$$\hat{\mathbf{y}} = X\hat{\boldsymbol{\beta}} = X\left(X^\top X\right)^{-1} X^\top \mathbf{y}.$$

Therefore the hat matrix is given by

$$H = X\left(X^\top X\right)^{-1} X^\top.$$

In the language of linear algebra, the hat matrix is the orthogonal projection onto the column space of the design matrix X. (Note that $\left(X^\top X\right)^{-1} X^\top$ is the pseudoinverse of X.)

The hat matrix corresponding to a linear model is symmetric and idempotent, that is, $H^2 = H$.

However, this is not always the case; in locally weighted scatterplot smoothing (LOESS), for example, the hat matrix is in general neither symmetric nor idempotent. The formula for the vector of residuals r can be expressed compactly using the hat matrix:

$$\mathbf{r} = \mathbf{y} - \hat{\mathbf{y}} = \mathbf{y} - H\mathbf{y} = (I - H)\mathbf{y}.$$

The covariance matrix of the residuals is therefore, by error propagation, equal to $(I-H)^{\top}\Sigma(I-H)$, where Σ is the covariance matrix of the errors (and by extension, the observations as well). For the case of linear models with independent and identically distributed errors in which $\Sigma = \sigma^2 I$, this reduces to $(I - H)\sigma^2$.

For linear models, the trace of the hat matrix is equal to the rank of X, which is the number of independent parameters of the linear model. For other models such as LOESS that are still linear in the observations y, the hat matrix can be used to define the effective degrees of freedom of the model.

The hat matrix has a number of useful algebraic properties. Practical applications of the hat matrix in regression analysis include leverage and Cook's distance, which are concerned with identifying observations which have a large effect on the results of a regression.

Correlated Errors

The above may be generalized to the case of correlated errors. Suppose that the covariance matrix of the errors is Σ. Then since

$$\hat{\beta} = \left(X^{\top}\Sigma^{-1}X\right)^{-1} X^{\top}\Sigma^{-1}\mathbf{y},$$

the hat matrix is thus

$$H = X\left(X^{\top}\Sigma^{-1}X\right)^{-1} X^{\top}\Sigma^{-1},$$

and again it may be seen that $H^2 = H$.

Blockwise Formula

Suppose the design matrix C can be decomposed by columns as $C = [A, B]$. Define the Hat operator as $H(X) = X\left(X^{\top}X\right)^{-1} X^{\top}$. Similarly, define the residual operator as $M(X) = I - H(X)$. Then the Hat matrix of can be decomposed as follows:

$$H(C) = H(A) + H(M(A)B)$$

There are a number of applications of such a partitioning. The classical application has a column of all ones, which allows one to analyze the effects of adding an intercept term to a regression. Another use is in the fixed effects model, where is a large sparse matrix of the dummy variables for the fixed effect terms.

One can use this partition to compute the hat matrix of without explicitly forming the matrix , which might be too large to fit into computer memory.

Hermitian Matrix

In mathematics, an Hermitian matrix (or self-adjoint matrix) is a square matrix with complex entries that is equal to its own conjugate transpose – that is, the element in the i-th row and j-th column is equal to the complex conjugate of the element in the j-th row and i-th column, for all indices i and j:

$$a_{ij} = \overline{a_{ji}}.$$

If the conjugate transpose of a matrix A is denoted by $A^{\dagger}$, then the Hermitian property can be written concisely as

$$A = A^{\dagger}.$$

Hermitian matrices can be understood as the complex extension of real symmetric matrices. Hermitian matrices are named after Charles Hermite, who demonstrated in 1855 that matrices of this form share a property with real symmetric matrices of having eigenvalues always real.

Examples

For example,

$$\begin{bmatrix} 3 & 2+i \\ 2-i & 1 \end{bmatrix}$$

Well-known families of Pauli matrices, Gell-Mann matrices and various generalizations are Hermitian. In theoretical physics such Hermitian matrices usually are multiplied by imaginary coefficients, which results in *skew-Hermitian* matrices.

Properties

The entries on the main diagonal (top left to bottom right) of any Hermitian matrix are necessarily real. A matrix that has only real entries is Hermitian if and only if it is a symmetric matrix, i.e., if it is symmetric with respect to the main diagonal. A real and symmetric matrix is simply a special case of a Hermitian matrix.

Every Hermitian matrix is a normal matrix, and the finite-dimensional spectral theorem applies. It says that any Hermitian matrix can be diagonalized by a unitary matrix, and that the resulting diagonal matrix has only real entries. This implies that all eigenvalues of a Hermitian matrix A are real, and that A has n linearly independent eigenvectors. Moreover, it is possible to find an orthonormal basis of C^n consisting of n eigenvectors of A.

The sum of any two Hermitian matrices is Hermitian, and the inverse of an invertible Hermitian matrix is Hermitian as well. However, the product of two Hermitian matrices A and B will only be Hermitian if they commute, i.e., if $AB = BA$. Thus A^n is Hermitian if A is Hermitian and n is an integer. The Hermitian complex n-by-n matrices do not form a vector space over the complex numbers, since the identity matrix I_n is Hermitian, but $i(I_n)$ is not. However the complex Hermitian matrices *do* form a vector space over the real numbers. In the $2n^2$ R dimensional vector space of complex $n \times n$ matrices, the complex Hermitian matrices form a subspace of dimension n^2. If E_{jk} denotes the n-by-n matrix with a 1 in the j,k position and zeros elsewhere, a basis can be described as follows:

$$E_{jj} \text{ for } 1 \le j \le n \ (n \text{ matrices})$$

together with the set of matrices of the form

$$E_{jk} + E_{kj} \text{ for } 1 \le j < k \le n \ ((n^2 - n)/2 \text{ matrices})$$

and the matrices

$$i(E_{jk} - E_{kj}) \text{ for } ((n^2 - n)/2 \text{ matrices})$$

where denotes the complex number $\sqrt{-1}$, known as the imaginary unit. If n orthonormal eigenvectors $u_1, \ldots, u_n$ of a Hermitian matrix are chosen and written as the columns of the matrix U, then one eigendecomposition of A is $A = U\Lambda U^\dagger$ where $UU^\dagger = I = U^\dagger U$ and therefore

$$A = \sum_j \lambda_j u_j u_j^\dagger$$

where λ_j are the eigenvalues on the diagonal of the diagonal matrix Λ.

Additional facts related to Hermitian matrices include:

- The sum of a square matrix and its conjugate transpose $(C + C^\dagger)$ is Hermitian.
- The difference of a square matrix and its conjugate transpose $(C - C^\dagger)$ is skew-Hermitian (also called antihermitian).
    - o This implies that commutator of two Hermitian matrices is skew-Hermitian.
- An arbitrary square matrix C can be written as the sum of a Hermitian matrix A and a skew-Hermitian matrix B:

$$C = A + B \quad \text{with} \quad A = \frac{1}{2}(C + C^\dagger) \quad \text{and} \quad B = \frac{1}{2}(C - C^\dagger).$$

- The determinant of a Hermitian matrix is real:

Proof:

$$\det(A) = \det(A^{\mathrm{T}}) \quad \Rightarrow \quad \det(A^\dagger) = \det(A)^*$$

Therefore if $A = A^\dagger \quad \Rightarrow \quad \det(A) = \det(A)^*$.

(Alternatively, the determinant is the product of the matrix's eigenvalues, and as mentioned before, the eigenvalues of a Hermitian matrix are real.)

Hessenberg Matrix

In linear algebra, a Hessenberg matrix is a special kind of square matrix, one that is "almost" triangular. To be exact, an upper Hessenberg matrix has zero entries below the first subdiagonal, and a lower Hessenberg matrix has zero entries above the first superdiagonal. They are named after Karl Hessenberg.

For example:

$$\begin{bmatrix} 1 & 4 & 2 & 3 \\ 3 & 4 & 1 & 7 \\ 0 & 2 & 3 & 4 \\ 0 & 0 & 1 & 3 \end{bmatrix}$$

is upper Hessenberg and

$$\begin{bmatrix} 1 & 2 & 0 & 0 \\ 5 & 2 & 3 & 0 \\ 3 & 4 & 3 & 7 \\ 5 & 6 & 1 & 1 \end{bmatrix}$$

is lower Hessenberg.

Computer Programming

Many linear algebra algorithms require significantly less computational effort when applied to triangular matrices, and this improvement often carries over to Hessenberg matrices as well. If the constraints of a linear algebra problem do not allow a general matrix to be conveniently reduced to a triangular one, reduction to Hessenberg form is often the next best thing. In fact, reduction of any matrix to a Hessenberg form can be achieved in a finite number of steps (for example, through Householder's algorithm of unitary similarity transforms).

Subsequent reduction of Hessenberg matrix to a triangular matrix can be achieved through iterative procedures, such as shifted QR-factorization.

In eigenvalue algorithms, the Hessenberg matrix can be further reduced to a triangular matrix through Shifted QR-factorization combined with deflation steps.

Reducing a general matrix to a Hessenberg matrix and then reducing further to a triangular matrix, instead of directly reducing a general matrix to a triangular matrix, often economizes the arithmetic involved in the QR algorithm for eigenvalue problems.

Properties

The product of a Hessenberg matrix with a triangular matrix is again Hessenberg. More precisely, if A is upper Hessenberg and T is upper triangular, then AT and TA are upper Hessenberg. A matrix that is both upper Hessenberg and lower Hessenberg is a tridiagonal matrix.

Hessian Matrix

In mathematics, the Hessian matrix (or simply the Hessian) is the square matrix of second-order partial derivatives of a function; that is, it describes the local curvature of a function of many variables.

The Hessian matrix was developed in the 19th century by the German mathematician Ludwig Otto Hesse and later named after him.

Hesse himself had used the term "functional determinants".

Given the real-valued function

$$f(x_1, x_2, \ldots, x_n),$$

if all second partial derivatives of f exist, then the Hessian matrix of f is the matrix

$$H(f)_{ij}(x) = D_i D_j f(x)$$

where $x = (x_1, x_2, \ldots, x_n)$ and D_i is the differentiation operator with respect to the ith argument and the Hessian becomes

$$H(f) = \begin{bmatrix} \frac{\partial^2 f}{\partial x_1^2} & \frac{\partial^2 f}{\partial x_1 \partial x_2} & \cdots & \frac{\partial^2 f}{\partial x_1 \partial x_n} \\ [2.2ex]\frac{\partial^2 f}{\partial x_2 \partial x_1} & \frac{\partial^2 f}{\partial x_2^2} & \cdots & \frac{\partial^2 f}{\partial x_2 \partial x_n} \\ [2.2ex]\vdots & \vdots & \ddots & \vdots \\ [2.2ex]\frac{\partial^2 f}{\partial x_n \partial x_1} & \frac{\partial^2 f}{\partial x_n \partial x_2} & \cdots & \frac{\partial^2 f}{\partial x_n^2} \end{bmatrix}.$$

Because f is often clear from context, $H(f)(x)$ is frequently shortened to simply $H(x)$.

The Hessian matrix is related to the Jacobian matrix by, $H(f)(x) = J(\nabla f)(x)$.

Some mathematicians define the Hessian as the determinant of the above matrix.

Hessian matrices are used in large-scale optimization problems within Newton-type methods because they are the coefficient of the quadratic term of a local Taylor expansion of a function. That is,

$$y = f(\mathbf{x}+\Delta\mathbf{x}) \approx f(\mathbf{x}) + J(\mathbf{x})\Delta\mathbf{x} + \frac{1}{2}\Delta\mathbf{x}^{\mathrm{T}} H(\mathbf{x})\Delta\mathbf{x}$$

where J is the Jacobian matrix, which is a vector (the gradient) for scalar-valued functions.

The full Hessian matrix can be difficult to compute in practice; in such situations, quasi-Newton algorithms have been developed that use approximations to the Hessian. The best-known quasi-Newton algorithm is the BFGS algorithm.

Mixed Derivatives and Symmetry of the Hessian

The mixed derivatives of f are the entries *off* the main diagonal in the Hessian. Assuming that they are continuous, the order of differentiation does not matter (Clairaut's theorem). For example,

$$\frac{\partial}{\partial x}\left(\frac{\partial f}{\partial y}\right) = \frac{\partial}{\partial y}\left(\frac{\partial f}{\partial x}\right).$$

This can also be written as:

$$f_{yx} = f_{xy}.$$

In a formal statement: if the second derivatives of f are all continuous in a neighbourhood D, then the Hessian of f is a symmetric matrix throughout D.

Critical Points and Discriminant

If the gradient of f (i.e. its derivative in the vector sense) is zero at some point x, then f has a *critical point* (or *stationary point*) at x. The determinant of the Hessian at x is then called the discriminant. If this determinant is zero then x is called a *degenerate critical point* of f, this is also called a *non-Morse critical point* of f. Otherwise it is non-degenerate, this is called a *Morse critical point* of f.

Second Derivative Test

The following test can be applied at a non-degenerate critical point x. If the Hessian is positive definite at x, then f attains a local minimum at x. If the Hessian is negative definite at x, then f attains a local maximum at x. If the Hessian has both positive and negative eigenvalues then x is a saddle point for f (this is true even if x is degenerate). Otherwise the test is inconclusive.

Note that for positive semidefinite and negative semidefinite Hessians the test is inconclusive (yet a conclusion can be made that f is locally convex or concave respectively). However, more can be said from the point of view of Morse theory.

In view of what has just been said, the second derivative test for functions of one and two variables is simple. In one variable, the Hessian contains just one second derivative; if it is positive then x is a local minimum, if it is negative then x is a local maximum; if it is zero then the test is inconclusive. In two variables, the determinant can be used, because the determinant is the product of the eigenvalues. If it is positive then the eigenvalues are both positive, or both negative. If it is negative then the two eigenvalues have different signs. If it is zero, then the second derivative test is inconclusive.

More generally, the second-order conditions that are sufficient for a local minimum or maximum can be expressed in terms of the sequence of principal (upper-leftmost) minors (determinants of sub-matrices) of the Hessian; these conditions are a special case of those given in the next section for bordered Hessians for constrained optimization—the case in which the number of constraints is zero.

Chapter 10

Centrosymmetric Matrix

In mathematics, especially in linear algebra and matrix theory, a centrosymmetric matrix is a matrix which is symmetric about its centre. More precisely, an $n \times n$ matrix $A = [A_{i,j}]$ is centrosymmetric when its entries satisfy

$$A_{i,j} = A_{n-i+1,n-j+1} \text{ for } 1 \leq i,j \leq n.$$

If J denotes the $n \times n$ matrix with 1 on the counterdiagonal and 0 elsewhere (that is, $J_{i,n+1-i} = 1$; $J_{i,j} = 0$ if $j \neq n+1-i$), then a matrix A is centrosymmetric if and only if $AJ = JA$. The matrix J is sometimes referred to as the exchange matrix.

- All 2×2 centrosymmetric matrices have the form $\begin{bmatrix} a & b \\ b & a \end{bmatrix}$.
- All 3×3 centrosymmetric matrices have the form $\begin{bmatrix} a & b & c \\ d & e & d \\ c & b & a \end{bmatrix}$.
- Symmetric Toeplitz matrices are centrosymmetric.

Algebraic Structure

If A and B are centrosymmetric matrices over a given field K, then so are $A+B$ and cA for any c in K. In addition, the matrix product AB is centrosymmetric, since $JAB = AJB = ABJ$.

Since the identity matrix is also centrosymmetric, it follows that the set of $n \times n$ centrosymmetric matrices over K is a subalgebra of the associative algebra of all $n \times n$ matrices.

Related Structures

An $n \times n$ matrix A is said to be *skew-centrosymmetric* if its entries satisfy $A_{i,j} = -A_{n-i+1,n-j+1}$ for $1 \leq i,j \leq n$. Equivalently, A is skew-centrosymmetric if $AJ = -JA$, where J is the exchange matrix defined above. The centrosymmetric relation $AJ = JA$ lends itself to a natural generalization, where J is replaced with an involutary matrix K (i.e., $K^2 = I$).

Symmetric centrosymmetric matrices are sometimes called bisymmetric matrices. When the ground field is the field of real numbers, it has been shown that bisymmetric matrices are precisely those symmetric matrices whose eigenvalues are the same up to sign after pre or post multiplication by the exchange matrix.

A similar result holds for Hermitian centrosymmetric and skew-centrosymmetric matrices.

Column Space

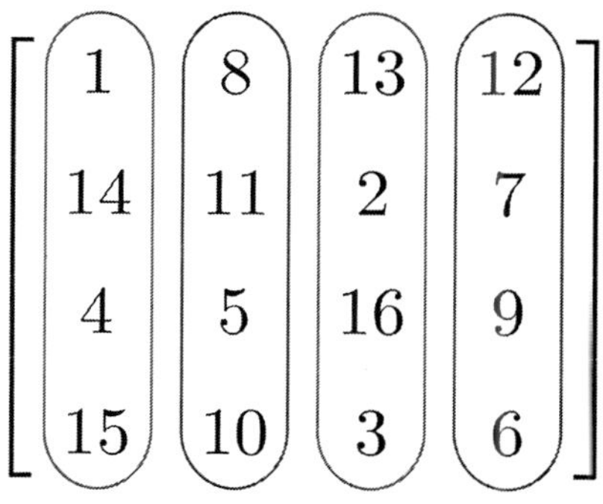

Figure: *The column vectors of a matrix.*

In linear algebra, the column space, C(A) of a matrix (sometimes called the range of a matrix) is the set of all possible linear combinations of its column vectors. The column space of an $m \times n$ matrix is a subspace of m-dimensional Euclidean space. The dimension of the column space is called the rank of the matrix. The column space of a matrix is the image or range of the corresponding matrix transformation.

Definition

Let A be an $m \times n$ matrix, with column vectors $v_1, v_2, \ldots, v_n$. A linear combination of these vectors is any vector of the form

$$c_1\mathbf{v}_1 + c_2\mathbf{v}_2 + \cdots + c_n\mathbf{v}_n,$$

where $c_1, c_2, ..., c_n$ are scalars. The set of all possible linear combinations of $v_1,...,v_n$ is called the column space of A. That is, the column space of A is the span of the vectors $v_1,...,v_n$.

Example

If $A = \begin{bmatrix} 1 & 0 \\ 0 & 1 \\ 2 & 0 \end{bmatrix}$, then the column vectors are $v_1 = (1, 0, 2)^T$ and $v_2 = (0, 1, 0)^T$. A linear combination of v_1 and v_2 is any vector of the form

$$c_1\begin{bmatrix} 1 \\ 0 \\ 2 \end{bmatrix} + c_2\begin{bmatrix} 0 \\ 1 \\ 0 \end{bmatrix} = \begin{bmatrix} c_1 \\ c_2 \\ 2c_1 \end{bmatrix}$$

The set of all such vectors is the column space of A. In this case, the column space is precisely the set of vectors (x, y, z) – R^3 satisfying the equation $z = 2x$ (using Cartesian coordinates, this set is a plane through the origin in three-dimensional space). Any linear combination of the column vectors of a matrix A can be written as the product of A with a column vector:

$$A\begin{bmatrix} x_1 \\ \vdots \\ x_n \end{bmatrix} = x_1\mathbf{v}_1 + \cdots + x_n\mathbf{v}_n$$

Therefore, the column space of A consists of all possible products Ax, for x – R^n. This is the same as the image (or range) of the corresponding matrix transformation.

Basis

The columns of A span the column space, but they may not form a basis if the column vectors are not linearly independent.

Fortunately, elementary row operations do not affect the dependence relations between the column vectors. This makes it possible to use row reduction to find a basis for the column space.

For example, consider the matrix

$$A = \begin{bmatrix} 1 & 3 & 1 & 4 \\ 2 & 7 & 3 & 9 \\ 1 & 5 & 3 & 1 \\ 1 & 2 & 0 & 8 \end{bmatrix}.$$

The columns of this matrix span the column space, but they may not be linearly independent, in which case some subset of them will form a basis. To find this basis, we reduce A to reduced row echelon form:

$$\begin{bmatrix} 1 & 3 & 1 & 4 \\ 2 & 7 & 3 & 9 \\ 1 & 5 & 3 & 1 \\ 1 & 2 & 0 & 8 \end{bmatrix} \sim \begin{bmatrix} 1 & 3 & 1 & 4 \\ 0 & 1 & 1 & 1 \\ 0 & 2 & 2 & -3 \\ 0 & -1 & -1 & 4 \end{bmatrix} \sim$$
$$\begin{bmatrix} 1 & 0 & -2 & 1 \\ 0 & 1 & 1 & 1 \\ 0 & 0 & 0 & -5 \\ 0 & 0 & 0 & 5 \end{bmatrix} \sim \begin{bmatrix} 1 & 0 & -2 & 0 \\ 0 & 1 & 1 & 0 \\ 0 & 0 & 0 & 1 \\ 0 & 0 & 0 & 0 \end{bmatrix}.$$

At this point, it is clear that the first, second, and fourth columns are linearly independent, while the third column is a linear combination of the first two. (Specifically, $v_3 = -2v_1 + v_2$.) Therefore, the first, second, and fourth columns of the original matrix are a basis for the column space:

$$\begin{bmatrix} 1 \\ 2 \\ 1 \\ 1 \end{bmatrix}, \begin{bmatrix} 3 \\ 7 \\ 5 \\ 2 \end{bmatrix}, \begin{bmatrix} 4 \\ 9 \\ 1 \\ 8 \end{bmatrix}.$$

Note that the independent columns of the reduced row echelon form are precisely the columns with pivots. This makes it possible to determine which columns are linearly independent by reducing only to echelon form.

The above algorithm can be used in general to find the dependence relations between any set of vectors, and to pick out a basis from any spanning set. A different algorithm for finding a basis from a spanning set is given in the row space article; finding a basis for the column space of A is equivalent to finding a basis for the row space of the transpose matrix A^T.

Dimension

The dimension of the column space is called the rank of the matrix. The rank is equal to the number of pivots in the reduced row echelon form, and is the maximum number of linearly independent columns that can be chosen from the matrix. For example, the 4×4 matrix in the example above has rank three.

Because the column space is the image of the corresponding matrix transformation, the rank of a matrix is the same as the dimension of the image. For example, the transformation $R^4 \rightarrow R^4$ described by the matrix above maps all of R^4 to some three-dimensional subspace.

The nullity of a matrix is the dimension of the null space, and is equal to the number of columns in the reduced row echelon form that do not have pivots. The rank and nullity of a matrix A with n columns are related by the equation:

$$\text{rank}(A) + \text{nullity}(A) = n.$$

This is known as the rank-nullity theorem.

Relation to the Left Null Space

The left null space of A is the set of all vectors x such that $x^T A = 0^T$. It is the same as the null space of the transpose of A. The left null space is the orthogonal complement to the column space of A.

This can be seen by writing the product of the matrix A^T and the vector x in terms of the dot product of vectors:

$$A^T \mathbf{x} = \begin{bmatrix} \mathbf{c}_1 \cdot \mathbf{x} \\ \mathbf{c}_2 \cdot \mathbf{x} \\ \vdots \\ \mathbf{c}_n \cdot \mathbf{x} \end{bmatrix},$$

where c_1, ..., c_n are the column vectors of A. Thus x = 0 if and only if x is orthogonal (perpendicular) to each of the column vectors of A.

It follows that the null space of is the orthogonal complement to the column space of A. For a matrix A, the column space, row space, null space, and left null space are sometimes referred to as the four fundamental subspaces.

Condition Number

In the field of numerical analysis, the condition number of a function with respect to an argument measures the asymptotically worst case of how much the function can change in proportion to small changes in the argument. The "function" is the solution of a problem and the "arguments" are the data in the problem.

A problem with a low condition number is said to be well-conditioned, while a problem with a high condition number is said to be ill-conditioned. The condition number is a property of the problem. Paired with the problem are any number of algorithms that can be used to solve the problem, that is, to calculate the solution. Some algorithms have a property called backward stability. In general, a backward stable algorithm can be expected to accurately solve well-conditioned problems. Numerical analysis textbooks give formulas for the condition numbers of problems and identify the backward stable algorithms. As a general rule of thumb, if the condition number $\kappa(A) = 10^k$, then you may lose up to k digits of accuracy on top of what would be lost to the numerical method due to loss of precision from arithmetic methods. However, the condition number does not give the exact value of the maximum inaccuracy that may occur in the algorithm. It generally just bounds it with an estimate (whose computed value depends on the choice of the norm to measure the inaccuracy).

Matrices

For example, the condition number associated with the linear equation $Ax = b$ gives a bound on how inaccurate the solution x will be after approximate solution. Note that this is before the effects of round-off error are taken into account; conditioning is a property of the matrix, not the algorithm or floating point accuracy of the computer used to solve the corresponding system. In particular, one should think of the condition number as being (very roughly) the rate at which the solution, x, will change with respect to a change in b. Thus, if the condition number is large, even a small error in b may cause a large error in x. On the other hand, if the condition number is small then the error in x will not be much bigger than the error in b.

The condition number is defined more precisely to be the maximum ratio of the relative error in x divided by the relative error in b.

Let e be the error in b. Assuming that A is a square matrix, the error in the solution $A^{-1}b$ is $A^{-1}e$. The ratio of the relative error in the solution to the relative error in b is

$$\frac{\|A^{-1}e\| / \|A^{-1}b\|}{\|e\| / \|b\|}.$$

This is easily transformed to

$$\left(\|A^{-1}e\| / \|e\|\right)\cdot\left(\|b\| / \|A^{-1}b\|\right).$$

The maximum value (for nonzero b and e) is easily seen to be the product of the two operator norms:

$$\kappa(A) = \|A\|\cdot\|A^{-1}\|.$$

The same definition is used for any consistent norm, i.e. one that satisfies

$$\kappa(A) \geq 1.$$

When the condition number is exactly one, then the algorithm may find an approximation of the solution with an arbitrary precision. However it does not mean that the algorithm will converge rapidly to this solution, just that it won't diverge arbitrarily because of inaccuracy on the source data (backward error), provided that the forward error introduced by the algorithm does not diverge as well because of accumulating intermediate rounding errors.

The condition number may also be infinite, in which case the algorithm will not reliably find a solution to the problem, not even a weak approximation of it (and not even its order of magnitude) with any reasonable and provable accuracy.

Of course, this definition depends on the choice of norm:

- If $\|\cdot\|$ is the norm (usually noted as $\|\cdot\|_2$) defined in the square-summable sequence space –!^{2} (which also matches the usual distance in a continuous and isotropic cartesian space), then

$$\kappa(A) = \frac{\sigma_{\max}(A)}{\sigma_{\min}(A)},$$

where $\sigma_{\max}(A)$ and $\lambda_{\min}(A)$ are maximal and minimal singular values of A respectively.

Hence

- o If A is normal then

$$\kappa(A) = \left|\frac{\lambda_{\max}(A)}{\lambda_{\min}(A)}\right|,$$

where $\lambda_{\max}(A)$ and $\lambda_{\min}(A)$ are maximal and minimal (by moduli) eigenvalues of respectively.

- o If is unitary then

$$\kappa(A) = 1.$$

This number arises so often in numerical linear algebra that it is given a name, the condition number of a matrix.

- If $\|\cdot\|$ is the norm (usually noted as ℓ^{∞}) defined in the sequence space –!⁻ of all bounded sequences (which also matches the non-linear distance measured as the maximum of distances measured on projections into the base subspaces, without requiring the space to be isotropic or even just linear, but only continuous, such norm being definable on all Banach spaces), and A is lower triangular non-singular (i.e., $\forall i, a_{ii} \neq 0$) then

$$\kappa(A) \geq \frac{\max_i(|a_{ii}|)}{\min_i(|a_{ii}|)}.$$

The condition number computed with this norm is generally larger than the condition number computed with square-summable sequences, but it can be evaluated more easily (and this is often the only measurable condition number, when the problem to solve involves a non-linear algebra, for example when approximating irrational and transcendental functions or numbers with numerical methods.) If the condition number is close to one, the matrix is well conditioned which means its inverse can be computed with good accuracy. If the condition number is large, then the matrix is said to be ill-conditioned. Practically, such a matrix is almost singular, and the computation of its inverse, or solution of a linear system of equations is prone to large numerical errors. A matrix that is not invertible has the condition number equal to infinity.

Other Contexts

Condition numbers can be defined for any function f mapping its data from some domain (e.g. an m-tuple of real numbers x) into some codomain [e.g. an n-tuple of real numbers $f(x)$], where both the domain and codomain are Banach spaces.

They express how sensitive that function is to small changes (or small errors) in its arguments. This is crucial in assessing the sensitivity and potential accuracy difficulties of numerous computational problems, for example polynomial root finding or computing eigenvalues.

The condition number of f at a point x (specifically, its relative condition number) is then defined to be the maximum ratio of the fractional change in $f(x)$ to any fractional change in x, in the limit where the change δx in x becomes infinitesimally small:

$$\lim_{\varepsilon \to 0^+} \sup_{\|\delta x\| \le \varepsilon} \left[\frac{\|f(x+\delta x) - f(x)\|}{\| f(x) \|} \Big/ \frac{\| \delta x \|}{\| x \|} \right],$$

where $\| \cdots \|$ is a norm on the domain/codomain of $f(x)$.

If f is differentiable, this is equivalent to:

$$\frac{\| J \|}{\| f(x) \| / \| x \|},$$

where J denotes the Jacobian matrix of partial derivatives of f and $\| J \|$ is the induced norm on the matrix.

Conductance (Graph)

In graph theory the conductance of a graph $G=(V,E)$ measures how "well-knit" the graph is: it controls how fast a random walk on G converges to a uniform distribution. The conductance of a graph is often called the Cheeger constant of a graph as the analog of its counterpart in spectral geometry.

Since electrical networks are intimately related to random walks with a long history in the usage of the term "conductance", this alternative name helps avoid possible confusion.

The conductance of a cut $(S,\bar{S})$ in a graph is defined as:

$$\varphi(S) = \frac{\sum_{i\in S, j\in \bar{S}} a_{ij}}{\min(a(S), a(\bar{S}))}$$

where the a_{ij} are the entries of the adjacency matrix for G, so that

$$a(S) = \sum_{i\in S}\sum_{j\in V} a_{ij}$$

is the total number (or weight) of the edges incident with S.

The conductance of the whole graph is the minimum conductance
over all the possible cuts:

$$\phi(G) = \min_{S\subseteq V} \varphi(S).$$

Equivalently, conductance of a graph is defined as follows:

$$\phi(G) = \min_{S\subseteq V} \varphi(S).\phi(G) := \min_{S\subseteq V; 0\leq a(S)\leq a(V)/2} \frac{\sum_{i\in S, j\in \bar{S}} a_{ij}}{a(S)}.$$

For a d-regular graph, the conductance is equal to the isoperimetric number divided by d.

Generalizations and Applications

In practical applications, one often considers the conductance only over a cut.

A common generalization of conductance is to handle the case of weights assigned to the edges: then the weights are added; if the weight is in the form of a resistance, then the reciprocal weights are added.

The notion of conductance underpins the study of percolation in physics and other applied areas; thus, for example, the permeability of petroleum through porous rock can be modeled in terms of the conductance of a graph, with weights given by pore sizes.

Markov Chains

For an ergodic reversible Markov chain with an underlying graph G, the conductance is a way to measure how hard it is to leave a small set of nodes. Formally, the conductance of a graph is defined as the minimum over all sets S of the capacity of S divided by the ergodic flow out of S. Alistair Sinclair showed that conductance is closely tied to mixing time in ergodic reversible Markov chains.

We can also view conductance in a more probabilistic way, as the minimal probability of leaving a small set of nodes given that we started in that set to begin with. Writing Φ_S for the conditional probability of leaving a set of nodes S given that we were in that set to begin with, the conductance is the minimal Φ_S over sets S that have a total stationary probability of at most 1/2.

Conductance is related to Markov chain mixing time in the reversible setting.

Conference Matrix

In mathematics, a conference matrix (also called a C-matrix) is a square matrix C with 0 on the diagonal and +1 and –1 off the diagonal, such that C^TC is a multiple of the identity matrix I. Thus, if the matrix has order n, $C^TC = (n-1)I$. Some authors use a more general definition, which requires there to be a single 0 in each row and column but not necessarily on the diagonal.

Conference matrices first arose in connection with a problem in telephony. They were first described by Vitold Belevitch who also gave them their name. Belevitch was interested in constructing ideal telephone conference networks from ideal transformers and discovered that such networks were represented by conference matrices, hence the name. Other applications are in statistics, and another is in elliptic geometry.

For $n > 1$, there are two kinds of conference matrix. Let us normalize C by, first (if the more general definition is used), rearranging the rows so that all the zeros are on the diagonal, and then negating any row or column whose first entry is negative.

(These operations do not change whether a matrix is a conference matrix.) Thus, a normalized conference matrix has all 1's in its first row and column, except for a 0 in the top left corner, and is 0 on the diagonal. Let S be the matrix that remains when the first row and column of C are removed.

Then either n is evenly even (a multiple of 4), and S is antisymmetric (as is the normalized C if its first row is negated), or n is oddly even (congruent to 2 modulo 4) and S is symmetric (as is the normalized C).

Symmetric Conference Matrices

If C is a symmetric conference matrix of order $n > 1$, then not only must n be congruent to 2 (mod 4) but also $n - 1$ must be a sum of two square integers; there is a clever proof by elementary matrix theory in van Lint and Seidel. n will always be the sum of two squares if $n - 1$ is a prime power.

Given a symmetric conference matrix, the matrix S can be viewed as the Seidel adjacency matrix of a graph.

The graph has $n - 1$ vertices, corresponding to the rows and columns of S, and two vertices are adjacent if the corresponding entry in S is negative. This graph is strongly regular of the type called (after the matrix) a conference graph.

The existence of conference matrices of orders n allowed by the above restrictions is known only for some values of n.

For instance, if $n = q + 1$ where q is a prime power congruent to 1 (mod 4), then the Paley graphs provide examples of symmetric conference matrices of order n, by taking S to be the Seidel matrix of the Paley graph.

The first few possible orders of a symmetric conference matrix are $n = 2, 6, 10, 14, 18$, (not 22, since 21 is not a sum of two squares), 26, 30, (not 34 since 33 is not a sum of two squares), 38, 42, 46, 50, 54, (not 58), 62 (sequence A000952 in OEIS); for every one of these, it is known that a symmetric conference matrix of that order exists. Order 66 seems to be an open problem.

Example

The essentially unique conference matrix of order 6 is given by

$$\begin{pmatrix} 0 & +1 & +1 & +1 & +1 & +1 \\ +1 & 0 & +1 & -1 & -1 & +1 \\ +1 & +1 & 0 & +1 & -1 & -1 \\ +1 & -1 & +1 & 0 & +1 & -1 \\ +1 & -1 & -1 & +1 & 0 & +1 \\ +1 & +1 & -1 & -1 & +1 & 0 \end{pmatrix},$$

all other conference matrices of order 6 are obtained from this one by flipping the signs of some row and/or column (and by taking per-mutations of rows and/or columns, according to the definition in use).

Antisymmetric Conference Matrices

Antisymmetric conference matrices can also be produced by the Paley construction. Let q be a prime power with residue 3 (mod 4). Then there is a Paley digraph of order q which leads to an antisymmetric conference matrix of order $n = q + 1$.

The matrix is obtained by taking for S the $q \times q$ matrix that has a +1 in position (i,j) and −1 in position (j,i) if there is an arc of the digraph from i to j, and zero diagonal. Then C constructed as above from S, but with the first row all negative, is an antisymmetric conference matrix. This construction solves only a small part of the problem of deciding for which evenly even numbers n there exist antisymmetric conference matrices of order n.

Generalizations

Sometimes a conference matrix of order n is just defined as a weighting matrix of the form $W(n, n-1)$, where $W(n,w)$ is said to be of weight $w>0$ and order n if it is a square matrix of size n with entries from $\{-1, 0, +1\}$ satisfying $W W^t = w I$. Using this definition, the zero element is no more required to be on the diagonal, but it is easy to see that still there must be exactly one zero element in each row and column. For example, the matrix

$$\begin{pmatrix} 1 & 0 & 1 & 1 \\ 0 & -1 & -1 & 1 \\ 1 & -1 & 0 & -1 \\ 1 & 1 & -1 & 0 \end{pmatrix}$$

would satisfy this relaxed definition, but not the more strict one requiring the zero elements to be on the diagonal.

Conformable Matrix

In mathematics, a matrix is conformable if its dimensions are suitable for defining some operation (*e.g.* addition, multiplication, etc.).

Examples

- In order to be conformable to addition, matrices need to have the same dimensions. Thus *A*, *B* and *C* all must have dimensions $m \times n$ in the equation $A + B = C$ for some fixed *m* and *n*.
- For matrix multiplication, consider the equation $AB = C$.

If *A* has dimensions $m \times n$, then *B* has to have dimensions $n \times p$ for some *p*, so that *C* will have dimensions $m \times p$.

Continuant (Mathematics)

In algebra, the continuant is a multivariate polynomial representing the determinant of a tridiagonal matrix and having applications in generalized continued fractions.

Definition

The *n*-th *continuant*, *K*(*n*), of a sequence a = $a_1,\ldots,a_n,\ldots$ is defined recursively by

$$K(0) = 1;$$

$$K(1) = a_1;$$

$$K(n) = a_n K(n-1) + K(n-2).$$

It may also be obtained by taking the sum of all possible products of $a_1,\ldots,a_n$ in which any pairs of consecutive terms are deleted.

An extended definition takes the continuant with respect to three sequences a, b and c, so that *K*(*n*) is a polynomial of $a_1,\ldots,a_n$, $b_1,\ldots,b_{n-1}$ and $c_1,\ldots,c_{n-1}$. In this case the recurrence relation becomes

$$K(0) = 1;$$

$$K(1) = a_1;$$

$$K(n) = a_n K(n-1) - b_{n-1}c_{n-1}K(n-2).$$

Since b_r and c_r enter into K only as a product $b_r c_r$ there is no loss of generality in assuming that the b_r are all equal to 1.

Applications

The simple continuant gives the value of a continued fraction of the form $[a_0, a_1, a_2, \ldots]$. The n-th convergent is

$$\frac{K(n+1,(a_0,\ldots,a_n))}{K(n,(a_1,\ldots,a_n))}.$$

The extended continuant is precisely the determinant of the tridiagonal matrix

$$\begin{pmatrix} a_1 & b_1 & 0 & 0 & \ldots & 0 & 0 \\ c_1 & a_2 & b_2 & 0 & \ldots & 0 & 0 \\ 0 & c_2 & a_3 & b_3 & \ldots & 0 & 0 \\ \vdots & \ddots & \ddots & \ddots & & \vdots & \vdots \\ 0 & 0 & 0 & 0 & \ldots & c_{n-1} & a_n \end{pmatrix}.$$

Copositive Matrix

In mathematics, specifically linear algebra, a real matrix A is copositive if

$$x^T A x \geq 0$$

for every nonnegative vector $x \geq 0$. The collection of all copositive matrices is a proper cone; it includes as a subset the collection of real positive-definite matrices.

Copositive matrices find applications in economics, operations research, and statistics.

Corner Transfer Matrix

In statistical mechanics, the corner transfer matrix describes the effect of adding a quadrant to a lattice.

Introduced by Rodney Baxter in 1968 as an extension of the Kramers-Wannier row-to-row transfer matrix, it provides a powerful method of studying lattice models. Calculations with corner transfer matrices led Baxter to the exact solution of the hard hexagon model in 1980.

Definition

Consider an IRF (interaction-round-a-face) model, i.e. a square lattice model with a spin σ_i assigned to each site i and interactions limited to spins around a common face. Let the total energy be given by

$$E = \sum_{\substack{all \\ faces}} \epsilon\left(\sigma_i, \sigma_j, \sigma_k, \sigma_l\right),$$

where for each face the surrounding sites i, j, k and l are arranged as follows:

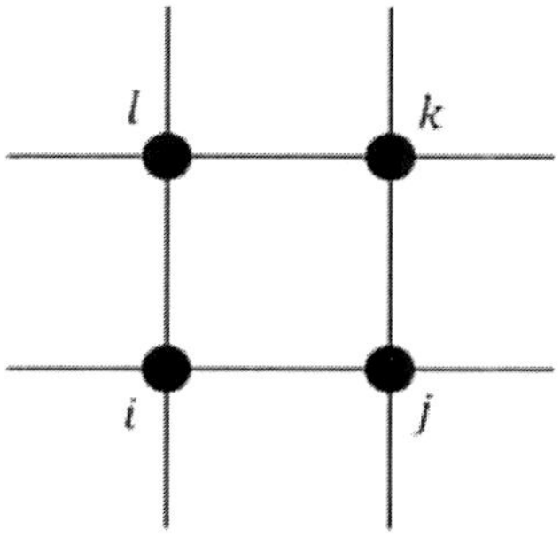

Figure: *For a lattice with* N *sites, the partition function is*

$$Z_N = \sum_{\substack{all \\ spins}} \prod_{\substack{all \\ faces}} w\left(\sigma_i, \sigma_j, \sigma_k, \sigma_l\right),$$

where the sum is over all possible spin configurations and w is the Boltzmann weight

$$w\left(\sigma_i, \sigma_j, \sigma_k, \sigma_l\right) = \exp\left(-\epsilon\left(\sigma_i, \sigma_j, \sigma_k, \sigma_l\right) / k_B T\right).$$

To simplify the notation, we use a ferromagnetic Ising-type lattice where each spin has the value +1 or −1, and the ground state is given by all spins up (i.e. the total energy is minimised when all spins on the lattice have the value +1).

We also assume the lattice has 4-fold rotational symmetry (up to boundary conditions) and is reflection-invariant. These simplifying assumptions are not crucial, and extending the definition to the general case is relatively straightforward.

Now consider the lattice quadrant shown below:

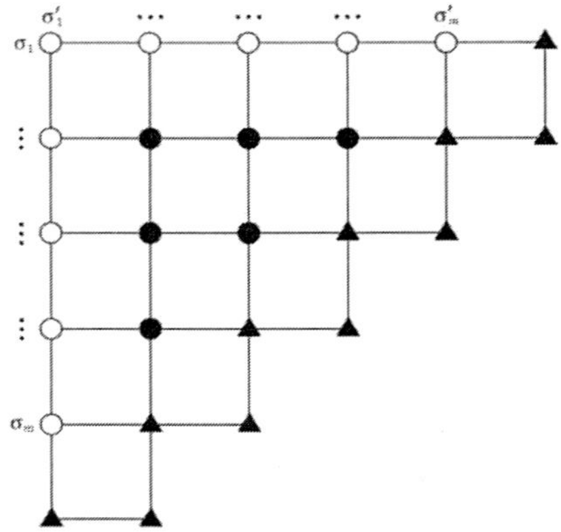

The outer boundary sites, marked by triangles, are assigned their ground state spins (+1 in this case).

The sites marked by open circles form the inner boundaries of the quadrant; their associated spin sets are labelled $\{\sigma_1,\ldots,\sigma_m\}$ and $\{\sigma'_1,\ldots,\sigma'_m\}$, where $\sigma_1 = \sigma'_1$. There are 2^m possible configurations for each inner boundary, so we define a $2^m \times 2^m$ matrix entry-wise by

$$A_{\sigma|\sigma'} = \delta(\sigma_1, \sigma_{1'}) \sum_{\substack{interior \\ spins}} \prod_{\substack{all \\ faces}} w(\sigma_i, \sigma_j, \sigma_k, \sigma_l).$$

The matrix A, then, is the corner transfer matrix for the given lattice quadrant. Since the outer boundary spins are fixed and the sum is over all interior spins, each entry of A is a function of the inner boundary spins.

The Kronecker delta in the expression ensures that $\sigma_1 = \sigma'_1$, so by ordering the configurations appropriately we may cast A as a block diagonal matrix:

$$\begin{array}{cc} & \begin{array}{ccc} \sigma_{1'} = +1 & & \sigma_{1'} = -1 \end{array} \\ A = & \left[\begin{array}{c|c} A_+ & 0 \\ \hline 0 & A_- \end{array}\right] \begin{array}{c} \sigma_1 = +1 \\ \\ \sigma_1 = -1 \end{array} \end{array}$$

Corner transfer matrices are related to the partition function in a simple way. In our simplified example, we construct the full lattice from four rotated copies of the lattice quadrant, where the inner boundary spin sets σ, σ', σ– and σ'– are allowed to differ:

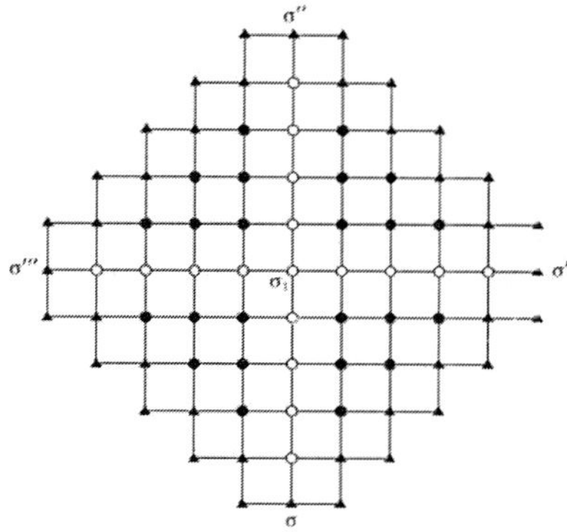

Figure: *The partition function is then written in terms of the corner transfer matrix* A *as*

$$Z_N = \sum_{\sigma,\sigma',\sigma'',\sigma'''} A_{\sigma|\sigma'} A_{\sigma'|\sigma''} A_{\sigma''|\sigma'''} A_{\sigma'''|\sigma} = \mathrm{tr} A^4.$$

Discussion

Recursion Relation

A corner transfer matrix A_2^m (defined for an $m \times m$ quadrant) may be expressed in terms of smaller corner transfer matrices A_2^{m-1} and A_2^{m-2} (defined for reduced $(m\text{-}1)\times(m\text{-}1)$ and $(m\text{-}2)\times(m\text{-}2)$ quadrants respectively). This recursion relation allows, in principle, the iterative calculation of the corner transfer matrix for any lattice quadrant of finite size. Like their row-to-row counterparts, corner transfer matrices may be factored into face transfer matrices, which correspond to adding a single face to the lattice. For the lattice quadrant given earlier, the face transfer matrices are of size $2^m \times 2^m$ and defined entry-wise by

$$\left(U_i\right)_{\sigma|\sigma'} = \delta\left(\sigma_1, \sigma_{1'}\right)\cdots\delta\left(\sigma_{i-1}, \sigma_{i-1'}\right)w$$
$$\left(\sigma_i, \sigma_{i+1}, \sigma_{i'}, \sigma_{i-1}\right)\delta\left(\sigma_{i+1}, \sigma_{i+1'}\right)\cdots\delta\left(\sigma_m, \sigma_{m'}\right),$$

where $2 \le i \le m+1$. Near the outer boundary, specifically, we have

$$\left(U_m\right)_{\sigma|\sigma'} = \delta\left(\sigma_1,\sigma_{1'}\right)\cdots\delta\left(\sigma_{m-1},\sigma_{m-1'}\right)w\left(\sigma_m,+1,\sigma_{m'},\sigma_{m-1}\right),$$

$$\left(U_{m+1}\right)_{\sigma|\sigma'} = \delta\left(\sigma_1,\sigma_{1'}\right)\cdots\delta\left(\sigma_m,\sigma_{m'}\right)w\left(+1,+1,+1,\sigma_m\right).$$

So the corner transfer matrix A factorises as

$$A = F_2\cdots F_{m+1},$$

where

$$F_j = U_{m+1}\cdots U_j.$$

Graphically, this corresponds to:

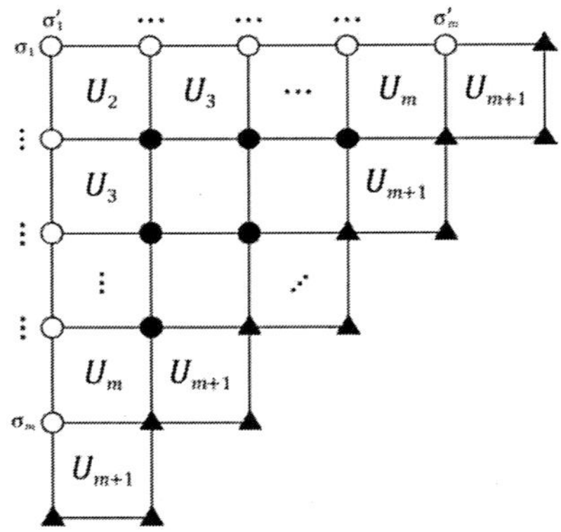

Figure: *We also require the* $2^m\times 2^m$ *matrices* A^* *and* A^{**}*, defined entry-wise by.*

$$\left(A^*\right)_{\sigma|\sigma'} = \delta\left(\sigma_1,\sigma_{1'}\right)A_{\sigma_2,\ldots,\sigma_m|\sigma_{2'},\ldots,\sigma_{m'}},$$

$$\left(A^{**}\right)_{\sigma|\sigma'} = \delta\left(\sigma_1,\sigma_{1'}\right)\delta\left(\sigma_2,\sigma_{2'}\right)A_{\sigma_3,\ldots,\sigma_m|\sigma_{3'},\ldots,\sigma_{m'}},$$

where the A matrices whose entries appear on the RHS are of size $2^{m-1}\times 2^{m-1}$ and $2^{m-2}\times 2^{m-2}$ respectively. This is more clearly written as

$$A^* = I_2 \otimes A_{2^{m-1}} = \begin{bmatrix} A & 0 \\ 0 & A \end{bmatrix},$$

$$A^{**} = I_2 \otimes I_2 \otimes A_{2^{m-2}} = \begin{bmatrix} A & 0 & 0 & 0 \\ 0 & A & 0 & 0 \\ 0 & 0 & A & 0 \\ 0 & 0 & 0 & A \end{bmatrix}.$$

Now from the definitions of A, A^*, A^{**}, U_i and F_j, we have

$$A^* = F_3 \cdots F_{m+1}, A^{**} = F_4 \cdots F_{m+1}$$

$$\Rightarrow A = F_2 A^*, A^* = F_3 A^{**}$$

$$\Rightarrow A = A^* \left(A^{**}\right)^{-1} U_2 A^*,$$

which gives the recursion relation for $A_2{}^m$ in terms of $A_2{}^{m-1}$ and $A_2{}^{m-2}$.

Diagonal Form

When using corner transfer matrices to perform calculations, it is both analytically and numerically convenient to work with their diagonal forms instead.

To facilitate this, the recursion relation may be rewritten directly in terms of the diagonal forms and eigenvector matrices of A, A^* and A^{**}.

Recalling that the lattice in our example is reflection-invariant, in the sense that

$$w\left(\sigma_i, \sigma_j, \sigma_k, \sigma_l\right) = w\left(\sigma_k, \sigma_j, \sigma_i, \sigma_l\right) = w\left(\sigma_i, \sigma_l, \sigma_k, \sigma_j\right),$$

we see that A is a symmetric matrix (i.e. it is diagonalisable by an orthogonal matrix). So we write

$$A = \alpha_m P A_d P^T,$$

where A_d is a diagonal matrix (normalised such that its numerically largest entry is 1), α_m is the largest eigenvalue of A, and $P^T P = I$. Likewise for A^* and A^{**}, we have

$$A^* = \alpha_{m-1} P^* A_d^* \left(P^*\right)^T,$$

$$A^{**} = \alpha_{m-2} P^{**} A_d^{**} \left(P^{**}\right)^T,$$

where $A_d{}^*$, $A_d{}^{**}$, P^* and P^{**} are defined in an analogous fashion to A^* and A^{**}, i.e. in terms of the smaller (normalised) diagonal forms and (orthogonal) eigenvector matrices of $A_2{}^{m-1}$ and $A_2{}^{m-2}$.

By substituting these diagonalisations into the recursion relation, we obtain

$$A_t = \kappa R A_d R^T,$$

where

$$\kappa = \frac{\alpha_m \alpha_{m-2}}{\alpha_{m-1}^2},$$

$$R = \left(P^*\right)^T P,$$

$$A_t = A_d^* \left(R^*\right)^T \left(A_d^{**}\right)^{-1} U_2 R^* A_d^*.$$

Now A_t is also symmetric, and may be calculated if A_d*, A_d** and R* are known; diagonalising A_t then yields its normalised diagonal form A_d, its largest eigenvalue κ, and its orthogonal eigenvector matrix R.

Applications

Spin Expectation Value

Corner transfer matrices (or their diagonal forms) may be used to calculate quantities such as the spin expectation value at a particular site deep inside the lattice.

For the full lattice given earlier, the spin expectation value at the central site is given by

$$\langle \sigma_1 \rangle = \frac{1}{Z_N} \sum_{\substack{all \\ spins}} \sigma_1 \prod_{\substack{all \\ faces}} w\left(\sigma_i, \sigma_j, \sigma_k, \sigma_l\right).$$

With the configurations ordered such that A is block diagonal as before, we may define a $2^m \times 2^m$ diagonal matrix

$$S = \begin{bmatrix} I & 0 \\ 0 & -I \end{bmatrix},$$

such that

$$\langle \sigma_1 \rangle = \frac{\mathrm{tr} S A^4}{\mathrm{tr} A^4} = \frac{\mathrm{tr}\, \alpha_m^4 P S A^4 P^T}{\mathrm{tr}\, \alpha_m^4 P A^4 P^T} = \frac{\mathrm{tr} S A_d^4}{\mathrm{tr} A_d^4}.$$

Partition Function Per Site

Another important quantity for lattice models is the partition function per site, evaluated in the thermodynamic limit and written as

$$\kappa = \lim_{N \to \infty} \left(Z_N \right)^{1/N}.$$

In our example, this reduces to

$$\kappa = \lim_{N \to \infty} \left(\alpha_m^4 \operatorname{tr} A_d^4 \right)^{1/N} \sim \alpha_m^{4/N},$$

since tr A_d^4 is a convergent sum as $m \to$ " and A_d becomes infinite-dimensional. Furthermore, the number of faces $2m(m+1)$ approaches the number of sites N in the thermodynamic limit, so we have

$$\alpha_m \sim \kappa^{m(m+1)/2},$$

which is consistent with the earlier equation giving κ as the largest eigenvalue for A_t. In other words, the partition function per site is given exactly by the diagonalised recursion relation for corner transfer matrices in the thermodynamic limit; this allows κ to be approximated via the iterative process of calculating A_d for a large lattice.

The matrices involved grow exponentially in size, however, and in actual numerical calculations they must be truncated at each step. One way of doing this is to keep the n largest eigenvalues at each step, for some fixed n. In most cases, the sequence of approximations obtained by taking n = 1,2,3,... converges rapidly, and to the exact value (for an exactly solvable model).

Covariance Matrix

In probability theory and statistics, a covariance matrix (also known as dispersion matrix or variance covariance matrix) is a matrix whose element in the i, j position is the covariance between the i th and j th elements of a random vector (that is, of a vector of random variables). Each element of the vector is a scalar random variable, either with a finite number of observed empirical values or with a finite or infinite number of potential values specified by a theoretical joint probability distribution of all the random variables.

Intuitively, the covariance matrix generalizes the notion of variance to multiple dimensions. As an example, the variation in a collection of random points in two-dimensional space cannot be characterized fully by a single number, nor would the variances in the x and y directions contain all of the necessary information; a 2×2 matrix would be necessary to fully characterize the two-dimensional variation. Analogous to the fact that it is necessary to build a Hessian matrix to fully describe the concavity of a multivariate function, a covariance matrix is necessary to fully describe the variation in a distribution. In this respect, a covariance matrix Σ of a dataset x with a joint probability density function is equals to the inverse of the Hessian matrix at the mean of dataset: $\Sigma_x = H(\overline{x})^{-1}$

Definition

Throughout this article, boldfaced unsubscripted X and Y are used to refer to random vectors, and unboldfaced subscripted X_i and Y_i are used to refer to random scalars. If the entries in the column vector

$$\mathbf{X} = \begin{bmatrix} X_1 \\ \vdots \\ X_n \end{bmatrix}$$

are random variables, each with finite variance, then the covariance matrix Σ is the matrix whose (i, j) entry is the covariance

$$\Sigma_{ij} = \mathrm{cov}(X_i, X_j) = \mathrm{E}\left[(X_i - \mu_i)(X_j - \mu_j)\right]$$

where

$$\mu_i = \mathrm{E}(X_i)$$

is the expected value of the ith entry in the vector X. In other words, we have

$$\Sigma = \begin{bmatrix} \mathrm{E}[(X_1-\mu_1)(X_1-\mu_1)] & \mathrm{E}[(X_1-\mu_1)(X_2-\mu_2)] & \cdots & \mathrm{E}[(X_1-\mu_1)(X_n-\mu_n)] \\ \mathrm{E}[(X_2-\mu_2)(X_1-\mu_1)] & \mathrm{E}[(X_2-\mu_2)(X_2-\mu_2)] & \cdots & \mathrm{E}[(X_2-\mu_2)(X_n-\mu_n)] \\ \vdots & \vdots & \ddots & \vdots \\ \mathrm{E}[(X_n-\mu_n)(X_1-\mu_1)] & \mathrm{E}[(X_n-\mu_n)(X_2-\mu_2)] & \cdots & \mathrm{E}[(X_n-\mu_n)(X_n-\mu_n)] \end{bmatrix}.$$

The inverse of this matrix, Σ^{-1}, is the inverse covariance matrix, also known as the concentration matrix or precision matrix. The elements of the precision matrix have an interpretation in terms of partial correlations and partial variances.

Generalization of the Variance

The definition above is equivalent to the matrix equality

$$\Sigma = \mathrm{E}\left[(\mathbf{X}-\mathrm{E}[\mathbf{X}])(\mathbf{X}-\mathrm{E}[\mathbf{X}])^{\mathrm{T}}\right]$$

This form can be seen as a generalization of the scalar-valued variance to higher dimensions. Recall that for a scalar-valued random variable X

$$\sigma^2 = \mathrm{var}(X) = \mathrm{E}[(X-\mathrm{E}(X))^2] = \mathrm{E}[(X-\mathrm{E}(X))\cdot(X-\mathrm{E}(X))].$$

Conflicting Nomenclatures and Notations

Nomenclatures differ. Some statisticians, following the probabilist William Feller, call this matrix the *variance* of the random vector X, because it is the natural generalization to higher dimensions of the 1-dimensional variance. Others call it the *covariance matrix*, because it is the matrix of covariances between the scalar components of the vector X. Thus

$$\mathrm{var}(\mathbf{X}) = \mathrm{cov}(\mathbf{X}) = \mathrm{E}\left[(\mathbf{X}-\mathrm{E}[\mathbf{X}])(\mathbf{X}-\mathrm{E}[\mathbf{X}])^{\mathrm{T}}\right].$$

However, the notation for the cross-covariance *between* two vectors is standard:

$$\mathrm{cov}(\mathbf{X},\mathbf{Y}) = \mathrm{E}\left[(\mathbf{X}-\mathrm{E}[\mathbf{X}])(\mathbf{Y}-\mathrm{E}[\mathbf{Y}])^{\mathrm{T}}\right].$$

The var notation is found in William Feller's two-volume book *An Introduction to Probability Theory and Its Applications*, but both forms are quite standard and there is no ambiguity between them. The matrix Σ is also often called the variance-covariance matrix since the diagonal terms are in fact variances.

Properties

For $\Sigma = \mathrm{E}\left[(\mathbf{X}-\mathrm{E}[\mathbf{X}])(\mathbf{X}-\mathrm{E}[\mathbf{X}])^{\mathrm{T}}\right]$ and $\mu = \mathrm{E}(\mathbf{X})$, where X is a random p-dimensional variable and Y a random q-dimensional variable, the following basic properties apply:

1. $\Sigma = E(\mathbf{XX}^T) - \mu\mu^T$
2. Σ is positive-semidefinite and symmetric.
3. $\text{cov}(\mathbf{AX} + \mathbf{a}) = \mathbf{A}\text{cov}(\mathbf{X})\mathbf{A}^T$
4. $\text{cov}(\mathbf{X}, \mathbf{Y}) = \text{cov}(\mathbf{Y}, \mathbf{X})^T$
5. $\text{cov}(\mathbf{X}_1 + \mathbf{X}_2, \mathbf{Y}) = \text{cov}(\mathbf{X}_1, \mathbf{Y}) + \text{cov}(\mathbf{X}_2, \mathbf{Y})$
6. If $p = q$, then
   $\text{var}(\mathbf{X} + \mathbf{Y}) = \text{var}(\mathbf{X}) + \text{cov}(\mathbf{X}, \mathbf{Y}) + \text{cov}(\mathbf{Y}, \mathbf{X}) + \text{var}(\mathbf{Y})$
7. $\text{cov}(\mathbf{AX}, \mathbf{B}^T\mathbf{Y}) = \mathbf{A}\text{cov}(\mathbf{X}, \mathbf{Y})\mathbf{B}$
8. If X and Y are independent, then $\text{cov}(\mathbf{X}, \mathbf{Y}) = 0$

where $\mathbf{X}, \mathbf{X}_1$ and $\mathbf{X}_2$ are random $p \times 1$ vectors, is a random $q \times 1$ vector, is $q \times 1$ vector, and are $q \times p$ matrices.

This covariance matrix is a useful tool in many different areas. From it a transformation matrix can be derived that allows one to completely decorrelate the data or, from a different point of view, to find an optimal basis for representing the data in a compact way. This is called principal components analysis (PCA) and the Karhunen-Loève transform (KL-transform).

Chapter 11

Permutation Matrix

In mathematics, a generalized permutation matrix (or monomial matrix) is a matrix with the same nonzero pattern as a permutation matrix, i.e. there is exactly one nonzero entry in each row and each column. Unlike a permutation matrix, where the nonzero entry must be 1, in a generalized permutation matrix the nonzero entry can be any nonzero value. An example of a generalized permutation matrix is

$$\begin{bmatrix} 0 & 0 & 3 & 0 \\ 0 & -2 & 0 & 0 \\ 1 & 0 & 0 & 0 \\ 0 & 0 & 0 & 1 \end{bmatrix}.$$

Structure

An invertible matrix A is a generalized permutation matrix if and only if it can be written as a product of an invertible diagonal matrix D and an (implicitly invertible) permutation matrix P: i.e.,

$$A = DP.$$

Group Structure

The set of $n \times n$ generalized permutation matrices with entries in a field F forms a subgroup of the general linear group $\mathrm{GL}(n,F)$, in which the group of nonsingular diagonal matrices $\Delta(n, F)$ forms a normal subgroup. Indeed, the generalized permutation matrices are the normalizer of the diagonal matrices, meaning that the generalized permutation matrices are the *largest* subgroup of GL in which diagonal matrices are normal.

The abstract group of generalized permutation matrices is **the wreath product of** $F^{\times}$ and S_n. Concretely, this means that it is the semidirect product of $\Delta(n, F)$ by the symmetric group S_n:

$$\Delta(n,F) \ltimes S_n,$$

where S_n acts by permuting coordinates and the diagonal matrices $\Delta(n, F)$ are isomorphic to the n-fold product $(F^{\times})^n$.

To be precise, the generalized permutation matrices are a (faithful) linear representation of this abstract wreath product: a realization of the abstract group as a subgroup of matrices.

Subgroups

- The subgroup where all entries are 1 is exactly the permutation matrices, which is isomorphic to the symmetric group.
- The subgroup where all entries are ±1 is the signed permutation matrices, which is the hyperoctahedral group.
- The subgroup where the entries are mth roots of unity μ_m is isomorphic to a generalized symmetric group.
- The subgroup of diagonal matrices is abelian, normal, and a maximal abelian subgroup. The quotient group is the symmetric group, and this construction is in fact the Weyl group of the general linear group: the diagonal matrices are a maximal torus in the general linear group (and are their own centralizer), the generalized permutation matrices are the normalizer of this torus, and the quotient, $N(T)/Z(T) = N(T)/T \cong S_n$ is the Weyl group.

Properties

- If a nonsingular matrix and its inverse are both nonnegative matrices (i.e. matrices with nonnegative entries), then the matrix is a generalized permutation matrix.

Generalizations

One can generalize further by allowing the entries to lie in a ring, rather than in a field. In that case if the non-zero entries are required to be units in the ring (invertible), one again obtains a group. On the other hand, if the non-zero entries are only required to be non-zero, but not necessarily invertible, this set of matrices forms a semigroup instead.

One may also schematically allow the non-zero entries to lie in a group *G,* with the understanding that matrix multiplication will only involve multiplying a single pair of group elements, not "adding" group elements. This is an abuse of notation, since element of matrices being multiplied must allow multiplication and addition, but is suggestive notion for the (formally correct) abstract group $G \wr S_n$ (the wreath product of the group G by the symmetric group).

Signed Permutation Group

A signed permutation matrix is a generalized permutation matrix whose nonzero entries are ±1, and are the integer generalized permutation matrices with integer inverse.

Properties

- It is the Coxeter group B_n, and has order $2^n n!$.
- It is the symmetry group of the hypercube and (dually) of the cross-polytope.
- Its index 2 subgroup of matrices with determinant 1 is the Coxeter group D_n and is the symmetry group of the demihypercube.
- It is a subgroup of the orthogonal group.

Monomial Representation

In mathematics, a linear representation ρ of a group G is a monomial representation if there is a finite-index subgroup H and a one-dimensional linear representation σ of H, such that ρ is equivalent to the induced representation

$$\mathrm{Ind}_H{}^G\sigma.$$

Alternatively, one may define it as a representation whose image is in the monomial matrices. Here for example G and H may be finite groups, so that *induced representation* has a classical sense. The monomial representation is only a little more complicated than the permutation representation of G on the cosets of H. It is necessary only to keep track of scalars coming from σ applied to elements of H.

Identity Matrix

In linear algebra, the identity matrix or unit matrix of size n is the $n\times n$ square matrix with ones on the main diagonal and zeros elsewhere. It is denoted by I_n, or simply by I if the size is immaterial or can be trivially deter-mined by the context. (In some fields, such as quantum mechanics, the identity matrix is denoted by a boldface one, 1; otherwise it is identical to I.)

$$I_1 = \begin{bmatrix}1\end{bmatrix}, I_2 = \begin{bmatrix}1 & 0\\ 0 & 1\end{bmatrix}, I_3 = \begin{bmatrix}1 & 0 & 0\\ 0 & 1 & 0\\ 0 & 0 & 1\end{bmatrix}, \cdots, I_n = \begin{bmatrix}1 & 0 & \cdots & 0\\ 0 & 1 & \cdots & 0\\ \vdots & \vdots & \ddots & \vdots\\ 0 & 0 & \cdots & 1\end{bmatrix}$$

Some mathematics books use U and E to represent the Identity Matrix (meaning "Unit Matrix" and "Elementary Matrix", or from the German "Einheitsmatrix", respectively), although I is considered more universal.

When A is $m\times n$, it is a property of matrix multiplication that

$$I_m A = AI_n = A.$$

In particular, the identity matrix serves as the unit of the ring of all $n\times n$ matrices, and as the identity element of the general linear group GL(n) consisting of all invertible $n\times n$ matrices. (The identity matrix itself is invertible, being its own inverse.)

Where $n\times n$ matrices are used to represent linear transformations from an n-dimensional vector space to itself, I_n represents the identity function, regardless of the basis. The ith column of an identity matrix is the unit vector e_i. It follows that the determinant of the identity matrix is 1 and the trace is n.

Using the notation that is sometimes used to concisely describe diagonal matrices, we can write:

$$I_n = \text{diag}(1,1,\ldots,1).$$

It can also be written using the Kronecker delta notation:

$$(I_n)_{ij} = \delta_{ij}.$$

The identity matrix also has the property that, when it is the product of two square matrices, the matrices can be said to be the inverse of one another.

The identity matrix of a given size is the only idempotent matrix of that size having full rank. That is, it is the only matrix such that (a) when multiplied by itself the result is itself, and (b) all of its rows, and all of its columns, are linearly independent.

The principal square root of an identity matrix is itself, and this is its only positive definite square root. However, every identity matrix with at least two rows and columns has an infinitude of symmetric square roots.

Nested Dissection

In numerical analysis, nested dissection is a divide and conquer heuristic for the solution of sparse symmetric systems of linear equations based on graph partitioning. Nested dissection was introduced by George (1973); the name was suggested by Garrett Birkhoff.

Nested dissection consists of the following steps:

- Form an undirected graph in which the vertices represent rows and columns of the system of linear equations, and an edge represents a nonzero entry in the sparse matrix representing the system.
- Recursively partition the graph into subgraphs using separators, small subsets of vertices the removal of which allows the graph to be partitioned into subgraphs with at most a constant fraction of the number of vertices.
- Perform Cholesky decomposition (a variant of Gaussian elimination for symmetric matrices), ordering the elimination of the variables by the recursive structure of the partition: each of the two subgraphs formed by removing the separator is eliminated first, and then the separator vertices are eliminated.

As a consequence of this algorithm, the fill-in (the set of nonzero matrix entries created in the Cholesky decomposition that are not part of the input matrix structure) is limited to at most the square of the separator size at each level of the recursive partition.

In particular, for planar graphs (frequently arising in the solution of sparse linear systems derived from two-dimensional finite element method meshes) the resulting matrix has O(n log n) nonzeros, due to the planar separator theorem guaranteeing separators of size O($-n$).

For arbitrary graphs there is a nested dissection that guarantees fill-in within a logarithmic factor of optimal, but this method is not guaranteed to achieve optimal fill-in and finding the optimal dissection is not a solved problem.

Pentadiagonal Matrix

In linear algebra, a pentadiagonal matrix is a matrix that is nearly diagonal; to be exact, it is a matrix in which the only nonzero entries are on the main diagonal, and the first two diagonals above and below it. So it is of the form

$$\begin{pmatrix} c_1 & d_1 & e_1 & 0 & \cdots & \cdots & 0 \\ b_1 & c_2 & d_2 & e_2 & \ddots & & \vdots \\ a_1 & b_2 & \ddots & \ddots & \ddots & \ddots & \vdots \\ 0 & a_2 & \ddots & \ddots & \ddots & e_{n-3} & 0 \\ \vdots & \ddots & \ddots & \ddots & \ddots & d_{n-2} & e_{n-2} \\ \vdots & & \ddots & a_{n-3} & b_{n-2} & c_{n-1} & d_{n-1} \\ 0 & \cdots & \cdots & 0 & a_{n-2} & b_{n-1} & c_n \end{pmatrix}.$$

It follows that a pentadiagonal matrix has at most $5n-6$ nonzero entries, where n is the size of the matrix. Hence, pentadiagonal matrices are sparse. This makes them useful in numerical analysis.

Shear Matrix

In mathematics, a shear matrix or transvection is an elementary matrix that represents the addition of a multiple of one row or column to another. Such a matrix may be derived by taking the identity matrix and replacing one of the zero elements with a non-zero value.

A typical shear matrix is shown below:

$$S = \begin{pmatrix} 1 & 0 & 0 & \lambda & 0 \\ 0 & 1 & 0 & 0 & 0 \\ 0 & 0 & 1 & 0 & 0 \\ 0 & 0 & 0 & 1 & 0 \\ 0 & 0 & 0 & 0 & 1 \end{pmatrix}.$$

The name *shear* reflects the fact that the matrix represents a shear transformation. Geometrically, such a transformation takes pairs of points in a linear space, that are purely axially separated along the axis whose row in the matrix contains the shear element, and effectively replaces those pairs by pairs whose separation is no longer purely axial but has two vector components. Thus, the shear axis is always an eigenvector of S. A shear parallel to the x axis results in $x' = x + \lambda y$ and $y' = y$. In matrix form:

$$\begin{pmatrix} x' \\ y' \end{pmatrix} = \begin{pmatrix} 1 & \lambda \\ 0 & 1 \end{pmatrix} \begin{pmatrix} x \\ y \end{pmatrix}.$$

Similarly, a shear parallel to the y axis has $x' = x$ and $y' = y + x$. In matrix form:

$$\begin{pmatrix} x' \\ y' \end{pmatrix} = \begin{pmatrix} 1 & 0 \\ \lambda & 1 \end{pmatrix} \begin{pmatrix} x \\ y \end{pmatrix}.$$

Clearly the determinant will always be 1, as no matter where the shear element is placed, it will be a member of a skew-diagonal that also contains zero elements (as all skew-diagonals have length at least two) hence its product will remain zero and won't contribute to the determinant. Thus every shear matrix has an inverse, and the inverse is simply a shear matrix with the shear element negated, representing a shear transformation in the opposite direction. In fact, this is part of an easily derived more general result: if S is a shear matrix with shear element, then S^n is a shear matrix whose shear element is simply n. Hence, raising a shear matrix to a power n multiplies its shear factor by n.

Properties

If S is an $n \times n$ shear matrix, then:

- S has rank n and therefore is invertible

- *1* is the only eigenvalue of S, so det $S = 1$ and trace $S = n$
- the eigenspace of S has *n-1* dimensions.
- S is asymmetric
- S may be made into a block matrix by at most 1 column interchange and 1 row interchange operation
- the area, volume, or any higher order interior capacity of a polytope is invariant under the shear transformation of the polytope's vertices.

Shift Matrix

In mathematics, a shift matrix is a binary matrix with ones only on the superdiagonal or subdiagonal, and zeroes elsewhere. A shift matrix U with ones on the superdiagonal is an upper shift matrix.

The alternative subdiagonal matrix L is unsurprisingly known as a lower shift matrix. The *(i,j)*:th component of U and L are

$$U_{ij} = \delta_{i+1,j}, \quad L_{ij} = \delta_{i,j+1},$$

where δ_{ij} is the Kronecker delta symbol.

For example, the *5×5* shift matrices are

$$U_5 = \begin{pmatrix} 0 & 1 & 0 & 0 & 0 \\ 0 & 0 & 1 & 0 & 0 \\ 0 & 0 & 0 & 1 & 0 \\ 0 & 0 & 0 & 0 & 1 \\ 0 & 0 & 0 & 0 & 0 \end{pmatrix} \quad L_5 = \begin{pmatrix} 0 & 0 & 0 & 0 & 0 \\ 1 & 0 & 0 & 0 & 0 \\ 0 & 1 & 0 & 0 & 0 \\ 0 & 0 & 1 & 0 & 0 \\ 0 & 0 & 0 & 1 & 0 \end{pmatrix}.$$

Clearly, the transpose of a lower shift matrix is an upper shift matrix and vice versa.

Premultiplying a matrix A by a lower shift matrix results in the elements of A being shifted downward by one position, with zeroes appearing in the top row.

Postmultiplication by a lower shift matrix results in a shift left. Similar operations involving an upper shift matrix result in the opposite shift.

Clearly all shift matrices are nilpotent; an n by n shift matrix S becomes the null matrix when raised to the power of its dimension n.

Properties

Let L and U be the n by n lower and upper shift matrices, respectively. The following properties hold for both U and L. Let us therefore only list the properties for U:

- $\det(U) = 0$
- $\operatorname{trace}(U) = 0$
- $\operatorname{rank}(U) = n-1$
- The characteristic polynomials of U is

  $$p_U(\lambda) = (-1)^n \lambda^n.$$

- $U^n = 0$. This follows from the previous property by the Cayley–Hamilton theorem.
- The permanent of U is 0.

The following properties show how U and L are related:

- $L^T = U$; $U^T = L$
- The null spaces of U and L are

  $$N(U) = \operatorname{span}\{(1,0,\ldots,0)^T\},$$

  $$N(L) = \operatorname{span}\{(0,\ldots,0,1)^T\}.$$

- The spectrum of U and L is $\{0\}$. The algebraic multiplicity of 0 is n, and its geometric multiplicity is 1. From the expressions for the null spaces, it follows that (up to a scaling) the only eigenvector for U is $(1,0,\ldots,0)^T$, and the only eigenvector for L is $(0,\ldots,0,1)^T$.
- For LU and UL we have

  $$UL = I - \operatorname{diag}(0,\ldots,0,1),$$

  $$LU = I - \operatorname{diag}(1,0,\ldots,0).$$

These matrices are both idempotent, symmetric, and have the same rank as U and L

- $L^{n-a}U^{n-a} + L^aU^a = U^{n-a}L^{n-a} + U^aL^a = I$ (the identity matrix), for any integer a between 0 and n inclusive.

Examples

$$S=\begin{pmatrix} 0 & 0 & 0 & 0 & 0 \\ 1 & 0 & 0 & 0 & 0 \\ 0 & 1 & 0 & 0 & 0 \\ 0 & 0 & 1 & 0 & 0 \\ 0 & 0 & 0 & 1 & 0 \end{pmatrix}; \quad A=\begin{pmatrix} 1 & 1 & 1 & 1 & 1 \\ 1 & 2 & 2 & 2 & 1 \\ 1 & 2 & 3 & 2 & 1 \\ 1 & 2 & 2 & 2 & 1 \\ 1 & 1 & 1 & 1 & 1 \end{pmatrix}.$$

$$\text{Then } SA=\begin{pmatrix} 0 & 0 & 0 & 0 & 0 \\ 1 & 1 & 1 & 1 & 1 \\ 1 & 2 & 2 & 2 & 1 \\ 1 & 2 & 3 & 2 & 1 \\ 1 & 2 & 2 & 2 & 1 \end{pmatrix}; \quad AS=\begin{pmatrix} 1 & 1 & 1 & 1 & 0 \\ 2 & 2 & 2 & 1 & 0 \\ 2 & 3 & 2 & 1 & 0 \\ 2 & 2 & 2 & 1 & 0 \\ 1 & 1 & 1 & 1 & 0 \end{pmatrix}.$$

Clearly there are many possible permutations. For example, $S^T AS$ is equal to the matrix A shifted up and left along the main diagonal.

$$S^T AS=\begin{pmatrix} 2 & 2 & 2 & 1 & 0 \\ 2 & 3 & 2 & 1 & 0 \\ 2 & 2 & 2 & 1 & 0 \\ 1 & 1 & 1 & 1 & 0 \\ 0 & 0 & 0 & 0 & 0 \end{pmatrix}.$$

Single-Entry Matrix

In mathematics a single-entry matrix is a matrix where a single element is one and the rest of the elements are zero, e.g.,

$$\mathbf{J}^{23}=\begin{bmatrix} 0 & 0 & 0 \\ 0 & 0 & 1 \\ 0 & 0 & 0 \end{bmatrix}.$$

It is a specific type of a sparse matrix.

Skyline Matrix

A skyline matrix is a form of a square, banded (and typically symmetric) matrix that reduces the storage requirement of a matrix due to its banded nature.

$$A_{skyline}(i,j) = \begin{cases} A_{original}(i, j-i) & \text{if } i \le j \le i + bandwidth \\ 0 & otherwise \end{cases}$$

For symmetrical matrices (with zero based indexing):

$$A_{skyline}(i,j) = \begin{cases} A_{original}(i, j-i) & \text{if } i - bandwidth_{left} < j \le i + bandwidth_{right} \\ 0 & otherwise \end{cases}$$

Examples

For a symmetric 6 by 6 matrix A with a bandwidth of 3 turns into the 6 by 3 matrix A_{skyline}.

Matrix A in original format:

$$\begin{bmatrix} A_{11} & A_{12} & A_{13} & 0 & \cdots & 0 \\ & A_{22} & A_{23} & A_{24} & \ddots & \vdots \\ & & A_{33} & A_{34} & A_{35} & 0 \\ & & & A_{44} & A_{45} & A_{46} \\ & sym & & & A_{55} & A_{56} \\ & & & & & A_{66} \end{bmatrix}$$

Matrix A_{skyline} in skyline format:

$$\begin{bmatrix} A_{11} & A_{12} & A_{13} \\ A_{22} & A_{23} & A_{24} \\ A_{33} & A_{34} & A_{35} \\ A_{44} & A_{45} & A_{46} \\ A_{55} & A_{56} & 0 \\ A_{66} & 0 & 0 \end{bmatrix}$$

For an unsymmetrical 6 by 6 matrix B with left and right band widths of 2 turns into the 6 by 3 matrix B_{skyline}.

Matrix B in original format:

$$\begin{bmatrix} B_{11} & B_{12} & 0 & \cdots & \cdots & 0 \\ B_{21} & B_{22} & B_{23} & \ddots & \ddots & \vdots \\ 0 & B_{32} & B_{33} & B_{34} & \ddots & \vdots \\ \vdots & \ddots & B_{43} & B_{44} & B_{45} & 0 \\ \vdots & \ddots & \ddots & B_{54} & B_{55} & B_{56} \\ 0 & \cdots & \cdots & 0 & B_{65} & B_{66} \end{bmatrix}$$

Matrix B_{skyline} in skyline format:

$$\begin{bmatrix} 0 & B_{11} & B_{12} \\ B_{21} & B_{22} & B_{23} \\ B_{32} & B_{33} & B_{34} \\ B_{43} & B_{44} & B_{45} \\ B_{54} & B_{55} & B_{56} \\ B_{65} & B_{66} & 0 \end{bmatrix}$$

Skyline Matrix Storage

Skyline matrix storage, or SKS, or a variable band matrix storage, or envelope storage scheme is a form of a sparse matrix storage format matrix that reduces the storage requirement of a matrix more than banded storage.

In banded storage, all entries within a fixed distance from the diagonal (called half-bandwidth) are stored. In column oriented skyline storage, only the entries from the first nonzero entry to the last nonzero entry in each column are stored. There is also row oriented skyline storage, and, for symmetric matrices, only one triangle is usually stored.

Skyline storage has become very popular in the finite element codes for structural mechanics, because the skyline is preserved by Cholesky decomposition (a method of solving systems of linear equations with a symmetric, positive-definite matrix; all fill-in falls within the skyline), and systems of equations from finite elements have a relatively small skyline.

In addition, the effort of coding skyline Cholesky is about same as for Cholesky for banded matrices.

Before storing a matrix in skyline format, the rows and columns are typically renumbered to reduce the size of the skyline (the number of nonzero entries stored) and to decrease the number of operations in the skyline Cholesky algorithm. The same heuristic renumbering algorithm that reduce the bandwidth are also used to reduce the skyline. The basic and one of the earliest algorithms to do that is reverse Cuthill–McKee algorithm.

However, skyline storage is not as popular for very large systems (many millions of equations) because skyline Cholesky is not so easily adapted for massively parallel computing, and general sparse methods, which store only the nonzero entries of the matrix, become more efficient for very large problems due to much less fill-in.

Sparse Matrix

In the subfield of numerical analysis, a sparse matrix is a matrix populated primarily with zeros (Stoer & Bulirsch 2002, p. 619). The term itself was coined by Harry M. Markowitz. If the majority of elements differ from zero then it is common to refer to the matrix as a dense matrix.

Conceptually, sparsity corresponds to systems which are loosely coupled. Consider a line of balls connected by springs from one to the next; this is a sparse system. By contrast, if the same line of balls had springs connecting each ball to all other balls, the system would be represented by a dense matrix. The concept of sparsity is useful in combinatorics and application areas such as network theory, which have a low density of significant data or connections.

Huge sparse matrices often appear in science or engineering when solving partial differential equations. When storing and manipulating sparse matrices on a computer, it is beneficial and often necessary to use specialized algorithms and data structures that take advantage of the sparse structure of the matrix.

Operations using standard dense matrix structures and algorithms are slow and consume large amounts of memory when applied to large sparse matrices. Sparse data is by nature easily compressed, and this compression almost always results in significantly less computer data storage usage. Indeed, some very large sparse matrices are infeasible to manipulate using standard dense algorithms.

Storing a Sparse Matrix

The native data structure for a matrix is a two-dimensional array. Each entry in the array represents an element $a_{i,j}$ of the matrix and can be accessed by the two indices i and j. For an $m \times n$ matrix, enough memory to store up to $(m \times n)$ entries to represent the matrix is needed.

Substantial memory requirement reductions can be realized by storing only the non-zero entries. Depending on the number and distribution of the non-zero entries, different data structures can be used and yield huge savings in memory when compared to a native approach.

Formats can be divided into two groups: (1) those that support efficient modification, and (2) those that support efficient matrix operations. The efficient modification group includes DOK (Dictionary of keys), LIL (List of lists), and COO (Coordinate list) and is typically used to construct the matrix. Once the matrix is constructed, it is typically converted to a format, such as CSR (Compressed Sparse Row) or CSC (Compressed Sparse Column), which is more efficient for matrix operations.

Dictionary of Keys (DOK)

DOK represents non-zero values as a dictionary mapping (row, column)-tuples to values. This format is good for incrementally constructing a sparse array, but poor for iterating over non-zero values in sorted order. One typically creates the matrix with this format, then converts to another format for processing

List of Lists (LIL)

LIL stores one list per row, where each entry stores a column index and value. Typically, these entries are kept sorted by column index for faster lookup. This is another format which is good for incremental matrix construction.

Coordinate List (COO)

COO stores a list of (row, column, value) tuples. Ideally, the entries are sorted (by row index, then column index) to improve random access times. This is another format which is good for incremental matrix construction.

Yale Format

The Yale Sparse Matrix Format stores an initial sparse $m \times n$ matrix, M, in row form using three one-dimensional arrays. Let NNZ denote the number of nonzero entries of M. The first array is A, which is of length NNZ, and holds all nonzero entries of M in left-to-right top-to-bottom (row-major) order.

The second array is IA, which is of length $m+1$ (i.e., one entry per row, plus one). IA(i) contains the index in A of the first nonzero element of row i. Row i of the original matrix extends from A(IA(i)) to A(IA(i+1)-1), i.e. from the start of one row to the last index before the start of the next. The third array, JA, contains the column index of each element of A, so it also is of length NNZ.

For example, the matrix

[1 2 0 0]
[0 3 9 0]
[0 1 4 0]

is a three-by-four matrix with six nonzero elements, so

A = [1 2 3 9 1 4]
IA = [0 2 4 6]
JA = [0 1 1 2 1 2]

In this case the Yale representation contains 16 entries, compared to only 12 in the original matrix. The Yale format saves on memory only when $NNZ < (m(n-1)-1)/2$

Another example, the matrix

[10 20 0 0 0 0]
[0 30 0 40 0 0]
[0 0 50 60 70 0]
[0 0 0 0 0 80]

is a four-by-six matrix (24 entries) with eight nonzero elements, so (spacing elements to be aligned with concerned data)

A = [10 20 30 40 50 60 70 80]
IA = [0 2 4 7 8]
JA = [0 1 1 3 2 3 4 5]

The whole is stored as 21 entries.

- IA splits the array A into rows: (10,20)(30,40)(50,60,70)(80);
- JA aligns values in columns: (10,20,...) (0,30,0,40,...) (0,0,50,60,70,0) (0,0,0,0,0,80).

(Note that in this format, the first value of IA will always be zero and the last will always be NNZ: these two cells may not be useful.)

Compressed Sparse Row (CSR or CRS)

CSR is effectively identical to the Yale Sparse Matrix format, except that the column array is normally stored ahead of the row index array. I.e. CSR is (val, col_ind, row_ptr), where val is an array of the (left-to-right, then top-to-bottom) non-zero values of the matrix; col_ind is the column indices corresponding to the values; and, row_ptr is the list of value indexes where each row starts.

The name is based on the fact that row index information is compressed relative to the COO format. One typically uses another format (LIL, DOK, COO) for construction. This format is efficient for arithmetic operations, row slicing, and matrix-vector products.

Compressed Sparse Column (CSC or CCS)

CSC is similar to CSR except that values are read first by column, a row index is stored for each value, and column pointers are stored. I.e. CSC is (val, row_ind, col_ptr), where val is an array of the (top-to-bottom, then left-to-right-bottom) non-zero values of the matrix; row_ind is the row indices corresponding to the values; and, col_ptr is the list of val indexes where each column starts. The name is based on the fact that column index information is compressed relative to the COO format. One typically uses another format (LIL, DOK, COO) for construction. This format is efficient for arithmetic operations, column slicing, and matrix-vector products. This is the traditional format for specifying a sparse matrix in MATLAB (via the sparse function).

Example

A bitmap image having only 2 colours, with one of them dominant (say a file that stores a handwritten signature) can be encoded as a sparse matrix that contains only row and column numbers for pixels with the non-dominant colour.

Band Matrix

An important special type of sparse matrices is band matrix, defined as follows. The *lower bandwidth* of a matrix A is the smallest number p such that the entry a_{ij} vanishes whenever $i > j + p$. Similarly, the *upper bandwidth* is the smallest p such that $a_{ij} = 0$ whenever $i < j - p$ (Golub & Van Loan 1996, §1.2.1). For example, a tridiagonal matrix has lower bandwidth 1 and upper bandwidth 1. As another example, the following sparse matrix has lower and upper bandwidth both equal to 3. Notice that zeros are represented with dots.

$$\begin{pmatrix} X & X & X & . & . & . & . \\ X & X & . & X & X & . & . \\ X & . & X & . & X & . & . \\ . & X & . & X & . & X & . \\ . & X & X & . & X & X & X \\ . & . & . & X & X & X & . \\ . & . & . & . & X & . & X \end{pmatrix}$$

Matrices with reasonably small upper and lower bandwidth are known as band matrices and often lend themselves to simpler algorithms than general sparse matrices; or one can sometimes apply dense matrix algorithms and gain efficiency simply by looping over a reduced number of indices. By rearranging the rows and columns of a matrix *A* it may be possible to obtain a matrix *A'* with a lower bandwidth. A number of algorithms are designed for bandwidth minimization.

Diagonal Matrix

A very efficient structure for an extreme case of band matrices, the diagonal matrix, is to store just the entries in the main diagonal as a one-dimensional array, so a diagonal $n \times n$ matrix requires only n entries.

Reducing Fill-in

The fill-in of a matrix are those entries which change from an initial zero to a non-zero value during the execution of an algorithm. To reduce the memory requirements and the number of arithmetic operations used during an algorithm it is useful to minimize the fill-in by switching rows and columns in the matrix. The symbolic Cholesky decomposition can be used to calculate the worst possible fill-in before doing the actual Cholesky decomposition.

There are other methods than the Cholesky decomposition in use. Orthogonalization methods (such as QR factorization) are common, for example, when solving problems by least squares methods. While the theoretical fill-in is still the same, in practical terms the "false non-zeros" can be different for different methods. And symbolic versions of those algorithms can be used in the same manner as the symbolic Cholesky to compute worst case fill-in.

Solving Sparse Matrix Equations

Both iterative and direct methods exist for sparse matrix solving. Iterative methods, such as conjugate gradient method and GMRES utilize fast computations of matrix-vector products Ax_i, where matrix A is sparse. The use of preconditioners can significantly accelerate convergence of such iterative methods.

Tridiagonal Matrix

In linear algebra, a tridiagonal matrix is a matrix that has nonzero elements only in the main diagonal, the first diagonal below this, and the first diagonal above the main diagonal.

For example, the following matrix is tridiagonal:

$$\begin{pmatrix} 1 & 4 & 0 & 0 \\ 3 & 4 & 1 & 0 \\ 0 & 2 & 3 & 4 \\ 0 & 0 & 1 & 3 \end{pmatrix}.$$

The determinant of a tridiagonal matrix is given by a continuant of its elements.

Determining an orthogonal transformation to tridiagonal form can be done with the Lanczos algorithm.

Properties

A tridiagonal matrix is of Hessenberg type; in particular, a tridiagonal matrix is a direct sum of p 1-by-1 and q 2-by-2 matrices such that p + q/2 = n — the dimension of the tridiagonal. Although a general tridiagonal matrix is not necessarily symmetric or Hermitian, many of those that arise when solving linear algebra problems have one of these properties.

Furthermore, if a real tridiagonal matrix A satisfies $a_{k,k+1}$ $a_{k+1,k} > 0$, so that the signs of its entries are symmetric, then it is similar to a Hermitian matrix, and hence, its eigenvalues are real.

The latter conclusion continues to hold if we replace the condition

$a_{k,k+1}\ a_{k+1,k} > 0$ by $a_{k,k+1}\ a_{k+1,k} \geq 0$.

The set of all $n \times n$ tridiagonal matrices forms a $3n$-2 dimensional vector space. Many linear algebra algorithms require significantly less computational effort when applied to diagonal matrices, and this improvement often carries over to tridiagonal matrices as well. For instance, the determinant of a tridiagonal matrix A of order n can be computed by the recursive formula for a continuant

$$\det[A]_{\{1,\ldots,n\}} = a_{n,n} \det[A]_{\{1,\ldots,n-1\}} - a_{n,n-1}a_{n-1,n} \det[A]_{\{1,\ldots,n-2\}},$$

where det $[A]_{\{1,\ldots,k\}}$ denotes the kth principal minor, that is, is the submatrix formed by the first k rows and columns of A. The cost of computing the determinant of a tridiagonal matrix using this formula is linear in n, while the cost is cubic for a general matrix.

Computer Programming

A transformation that reduces a general matrix to Hessenberg form will reduce a Hermitian matrix to *tridiagonal* form. So, many eigenvalue algorithms, when applied to a Hermitian matrix, reduce the input Hermitian matrix to tridiagonal form as a first step. A *tridiagonal matrix* can also be stored more efficiently than a general matrix by using a special storage scheme. For instance, the LAPACK Fortran package stores an unsymmetric tridiagonal matrix of order n in three one-dimensional arrays, one of length n containing the diagonal elements, and two of length n – 1 containing the subdiagonal and superdiagonal elements.

A system of tridiagonal matrix $Ax = b$, for $b \in \mathbb{R}^n$ can be solved by a specific algorithm called Tridiagonal matrix algorithm, requiring *O(n)* operations (Golub and Van Loan).

Zero Matrix

In mathematics, particularly linear algebra, a zero matrix or null matrix is a matrix with all its entries being zero. Some examples of zero matrices are

$$0_{1,1} = \begin{bmatrix} 0 \end{bmatrix}, 0_{2,2} = \begin{bmatrix} 0 & 0 \\ 0 & 0 \end{bmatrix}, 0_{2,3} = \begin{bmatrix} 0 & 0 & 0 \\ 0 & 0 & 0 \end{bmatrix}.$$

The set of $m \times n$ matrices with entries in a ring K forms a ring $K_{m,n}$. The zero matrix $0_{K_{m,n}}$ in is the matrix with all entries equal to , where is the additive identity in K.

$$0_{K_{m,n}} = \begin{bmatrix} 0_K & 0_K & \cdots & 0_K \\ 0_K & 0_K & \cdots & 0_K \\ \vdots & \vdots & \ddots & \vdots \\ 0_K & 0_K & \cdots & 0_K \end{bmatrix}_{m \times n}$$

The zero matrix is the additive identity in $K_{m,n}$. That is, for all $A \in K_{m,n}$ it satisfies

$$0_{K_{m,n}} + A = A + 0_{K_{m,n}} = A.$$

There is exactly one zero matrix of any given size $m \times n$ having entries in a given ring, so when the context is clear one often refers to *the* zero matrix.

In general the zero element of a ring is unique and typically denoted as 0 without any subscript indicating the parent ring. Hence the examples above represent zero matrices over any ring.

The zero matrix represents the linear transformation sending all vectors to the zero vector.

A matrix where just a single element is one and the rest are zero may be called a single-entry matrix.

Perfect Matrix

In mathematics, a perfect matrix is an m-by-n binary matrix that has no possible k-by-k submatrix K that satisfies the following conditions:

- $k > 3$
- The row and column sums of K are each equal to b, where $b \geq 2$
- There exists no row of the $(m - k)$-by-k submatrix formed by the rows not included in K with a row sum greater than b.

The following is an example of a K submatrix where $k = 5$ and $b = 2$:

$$\begin{bmatrix} 1 & 1 & 0 & 0 & 0 \\ 0 & 1 & 1 & 0 & 0 \\ 0 & 0 & 1 & 1 & 0 \\ 0 & 0 & 0 & 1 & 1 \\ 1 & 0 & 0 & 0 & 1 \end{bmatrix}.$$

Adjacency Matrix

In mathematics and computer science, an adjacency matrix is a means of representing which vertices (or nodes) of a graph are adjacent to which other vertices. Another matrix representation for a graph is the incidence matrix.

Specifically, the adjacency matrix of a finite graph G on n vertices is the $n \times n$ matrix where the non-diagonal entry a_{ij} is the number of edges from vertex i to vertex j, and the diagonal entry a_{ii}, depending on the convention, is either once or twice the number of edges (loops) from vertex i to itself. Undirected graphs often use the latter convention of counting loops twice, whereas directed graphs typically use the former convention.

There exists a unique adjacency matrix for each isomorphism class of graphs (up to permuting rows and columns), and it is not the adjacency matrix of any other isomorphism class of graphs. In the special case of a finite simple graph, the adjacency matrix is a (0,1)-matrix with zeros on its diagonal. If the graph is undirected, the adjacency matrix is symmetric.

The relationship between a graph and the eigenvalues and eigenvectors of its adjacency matrix is studied in spectral graph theory.

Algebraic Riccati Equation

The algebraic Riccati equation is either of the following matrix equations:

the continuous time algebraic Riccati equation (CARE):

$$A^T X + XA - XBR^{-1}B^T X + Q = 0$$

or the discrete time algebraic Riccati equation (DARE):

$$X = A^T XA - (A^T XB)(R + B^T XB)^{-1}(B^T XA) + Q.$$

X is the unknown n by n symmetric matrix and A, B, Q, R are known real coefficient matrices.

The name Riccati is given to the CARE equation by analogy to the Riccati differential equation: the unknown appears linearly and in a quadratic term (but no higher-order terms).

The DARE arises in place of the CARE when studying discrete time systems; it is not obviously related to the differential equation studied by Riccati.

The algebraic Riccati equation determines the solution of the infinite horizon time-invariant Linear-Quadratic Regulator problem (LQR) as well as that of the infinite horizon time-invariant Linear-Quadratic-Gaussian control problem (LQG). These are two of the most fundamental problems in control theory.

A solution to the algebraic Riccati equation can be obtained by matrix factorizations or by iterating on the Riccati equation.

Alternating Sign Matrix

In mathematics, an alternating sign matrix is a square matrix of 0s, 1s, and “1s such that the sum of each row and column is 1 and the nonzero entries in each row and column alternate in sign. These matrices arise naturally when using Dodgson condensation to compute a determinant. They are also closely related to the six vertex model with domain wall boundary conditions from statistical mechanics. They were first defined by William Mills, David Robbins, and Howard Rumsey in the former context.

For example, the permutation matrices are alternating sign matrices, as is

$$\begin{bmatrix} 0 & 0 & 1 & 0 \\ 1 & 0 & 0 & 0 \\ 0 & 1 & -1 & 1 \\ 0 & 0 & 1 & 0 \end{bmatrix}.$$

The *alternating sign matrix conjecture* states that the number of $n \times n$ alternating sign matrices is

$$\prod_{k=0}^{n-1} \frac{(3k+1)!}{(n+k)!} = \frac{1!4!7!\cdots(3n-2)!}{n!(n+1)!\cdots(2n-1)!}.$$

This conjecture was first proved by Doron Zeilberger in 1992. In 1995, Greg Kuperberg gave a short proof based on the Yang-Baxter equation for the six vertex model with domain wall boundary conditions, that uses a determinant due to Anatoli Izergin, which solves recurrence relations due to Vladimir Korepin.

Razumov–Stroganov Conjecture

In 2001 A.Razumov and Y.Stroganov conjectured a connection between O(1) loop model, fully packaged loop model (FPL) and ASMs. This conjecture was proved in 2010 by Cantini and Sportiello.

Arrowhead Matrix

In the mathematical field of linear algebra, an arrowhead matrix is a square matrix containing zeros in all entries except for the first row, first column, and main diagonal. In other words, the matrix has the form

$$\begin{pmatrix} * & * & * & * & * \\ * & * & 0 & 0 & 0 \\ * & 0 & * & 0 & 0 \\ * & 0 & 0 & * & 0 \\ * & 0 & 0 & 0 & * \end{pmatrix}.$$

Arrowhead matrices are used in some algorithms for finding of eigenvalues and eigenvectors.

Augmented Matrix

In linear algebra, an augmented matrix is a matrix obtained by appending the columns of two given matrices, usually for the purpose of performing the same elementary row operations on each of the given matrices.

Given the matrices A and B, where:

$$A = \begin{bmatrix} 1 & 3 & 2 \\ 2 & 0 & 1 \\ 5 & 2 & 2 \end{bmatrix}, \quad B = \begin{bmatrix} 4 \\ 3 \\ 1 \end{bmatrix}.$$

Then, the augmented matrix $(A \mid B)$ is written as:

$$(A \mid B) = \begin{bmatrix} 1 & 3 & 2 & 4 \\ 2 & 0 & 1 & 3 \\ 5 & 2 & 2 & 1 \end{bmatrix}.$$

This is useful when solving systems of linear equations.

The number of solutions to a system of linear equations depends only on the rank of the matrix representing the system and the corresponding augmented matrix.

An augmented matrix may also be used to find the inverse of a matrix by combining it with the identity matrix.

Examples

$$(C \mid I) = \begin{bmatrix} 1 & 3 & 1 & 0 \\ -5 & 0 & 0 & 1 \end{bmatrix}$$

$$(I \mid C^{-1}) = \begin{bmatrix} 1 & 0 & 0 & -\frac{1}{5} \\ 0 & 1 & \frac{1}{3} & \frac{1}{15} \end{bmatrix}$$

Solution of a Linear System

As used in linear algebra, an augmented matrix is used to represent the coefficients and the solution vector of each equation set. For the set of equations:

$$\begin{aligned} x_1 + 2x_2 + 3x_3 &= 0 \\ 3x_1 + 4x_2 + 7x_3 &= 2 \\ 6x_1 + 5x_2 + 9x_3 &= 11 \end{aligned}$$

the coefficients and constant terms give the matrices

$$A = \begin{bmatrix} 1 & 2 & 3 \\ 3 & 4 & 7 \\ 6 & 5 & 9 \end{bmatrix}, \quad B = \begin{bmatrix} 0 \\ 2 \\ 11 \end{bmatrix}.$$

Giving the augmented matrix:

$$(A \mid B) = \begin{bmatrix} 1 & 2 & 3 & 0 \\ 3 & 4 & 7 & 2 \\ 6 & 5 & 9 & 11 \end{bmatrix}$$

Balanced Matrix

In mathematics, a balanced matrix B is an integer matrix that does not contain any odd order 2-cycle submatrices (submatrix of order n where n is odd and the row and column sums equal 2). Balanced matrices are important in linear programs such as the set partitioning problem, as they are naturally integer. Totally unimodular matrices are a subset of balanced matrices, and balanced matrices are a subset of perfect matrices, therefore any matrix that is totally unimodular is also balanced, however a balanced matrix may not necessarily be totally unimodular.

The following matrix is a 3 order 2-cycle submatrix:

$$\begin{bmatrix} 1 & 0 & 1 \\ 1 & 1 & 0 \\ 0 & 1 & 1 \end{bmatrix}$$

Birkhoff Factorization

In mathematics, Birkhoff factorization or Birkhoff decomposition, introduced by Birkhoff (1909), is the factorization of an invertible matrix M with coefficients that are Laurent polynomials in z into a product $M = M^+M^0M^-$, where M^+ has entries that are polynomials in z, M^0 is diagonal, and M^- has entries that are polynomials in z^{-1}.

There are several variations where the general linear group is replaced by some other reductive algebraic group, due to Grothendieck (1957). Birkhoff factorization implies the Birkhoff–Grothendieck theorem of Grothendieck (1957) that vector bundles over the projective line are sums of line bundles.

Birkhoff factorization follows from the Bruhat decomposition for affine algebraic groups (or loop groups), and conversely the Bruhat decomposition for the affine general linear group follows from Birkhoff factorization together with the Bruhat decomposition for the ordinary general linear group.

Birkhoff Polytope

The Birkhoff polytope B_n, also called the assignment polytope, the polytope of doubly stochastic matrices, or the perfect matching polytope of the complete bipartite graph $K_{n,n}$, is the convex polytope in R^N (where $N = n^2$) whose points are the doubly stochastic matrices, i.e., the $n \times n$ matrices whose entries are non-negative real numbers and whose rows and columns each add up to 1.

Properties

Vertices: The Birkhoff polytope has $n!$ vertices, one for each permutation on n items.

This follows from the Birkhoff–von Neumann theorem states that the extreme points of the Birkhoff polytope are the permutation matrices, and therefore that any doubly stochastic matrix may be represented as a convex combination of permutation matrices; this was stated in a 1946 paper by Garrett Birkhoff, but equivalent results in the languages of projective configurations and of regular bipartite graph matchings, respectively, were shown much earlier in 1894 in Ernst Steinitz's thesis and in 1916 by Dénes Kõnig.

Edges

The edges of the Birkhoff polytope correspond to pairs of permutations differing by a cycle:

$$(\sigma,\omega) \text{ such that } \sigma^{-1}\omega \text{ is a cycle.}$$

This implies that the graph of B_n is a Cayley graph of the symmetric group S_n. This also implies that the graph of B_3 is a complete graph K_6, and thus B_3 is a neighbourly polytope.

Facets

The Birkhoff polytope lies within an $(n^2 - 2n + 1)$-dimensional affine subspace of the n^2-dimensional space of all $n \times n$ matrices: this subspace is determined by the linear equality constraints that the sum of each row and of each column be one. Within this subspace, it is defined by n^2 linear inequalities, one for each coordinate of the matrix, specifying that the coordinate be non-negative. Therefore, it has exactly n^2 facets.

Symmetries

The Birkhoff polytope B_n is both vertex-transitive and face-transitive (i.e. the dual polytope is vertex-transitive). It is not regular for $n>2$.

Volume

An outstanding problem is to find the volume of the Birkhoff polytopes. This has been done for $n \leq 10$. It is known to be equal to the volume of a polytope associated with standard Young tableaux. A combinatorial formula for all n was given in 2007. The following asymptotic formula was found by Rodney Canfield and Brendan McKay:

$$\mathrm{vol}(B_n)=\exp\left(-(n-1)^2 \ln n+n^2-(n-\frac{1}{2})\ln(2\pi)+\frac{1}{3}+o(1)\right).$$

Generalizations

- The Birkhoff polytope is a special case of the transportation polytope, a polytope of nonnegative rectangular matrices with given row and column sums. The integer points in these polytopes are called contingency tables; they play an important role in Bayesian statistics.
- The Birkhoff polytope is a special case of the matching polytope, defined as a convex hull of the perfect matchings in a finite graph. The description of facets in this generality was given by Jack Edmonds (1965), and is related to Edmonds's matching algorithm.

Bisymmetric Matrix

In mathematics, a bisymmetric matrix is a square matrix that is symmetric about both of its main diagonals. More precisely, an $n \times n$ matrix A is bisymmetric if it satisfies both $A = A^T$ and $AJ = JA$ where J is the $n \times n$ exchange matrix.

For example:

$$\begin{bmatrix} a & b & c & d & e \\ b & f & g & h & d \\ c & g & i & g & c \\ d & h & g & f & b \\ e & d & c & b & a \end{bmatrix}.$$

Properties

Bisymmetric matrices are both symmetric centrosymmetric and symmetric persymmetric. It has been shown that real-valued bisymmetric matrices are precisely those symmetric matrices whose eigenvalues are the same up to sign after pre or post multiplication by the exchange matrix.

The product of two bisymmetric matrices results in a centrosymetric matrix

Block Reflector

"A block reflector is an orthogonal, symmetric matrix that reverses a subspace whose dimension may be greater than one."

It is built out of many elementary reflectors.

It is also referred to as a triangular factor, and is a triangular matrix and they are used in the Householder transformation.

A reflector Q belonging to $\mathcal{M}_n(\mathbb{R})$ can be written in the form : $Q = I - auu^T$ where I is the identity matrix for $\mathcal{M}_n(\mathbb{R})$, a is a scalar and u belongs to R^n.

LAPACK Routines

Here are some of the LAPACK routines that apply to block reflectors

- "*larft" forms the triangular vector T of a block reflector H=I-VTVH.
- "*larzb" applies a block reflector or its transpose/ conjugate transpose as returned by "*tzrzf" to a general matrix.
- "*larzt" forms the triangular vector T of a block reflector H=I-VTVH as returned by "*tzrzf".
- "*larfb" applies a block reflector or its transpose/ conjugate transpose to a general rectangular matrix.

BLOSUM

The BLOSUM (BLOcks of Amino Acid SUbstitution Matrix) matrix is a substitution matrix used for sequence alignment of proteins. BLOSUM matrices are used to score alignments between evolutionarily divergent protein sequences.

They are based on local alignments. BLOSUM matrices were first introduced in a paper by Henikoff and Henikoff.

They scanned the BLOCKS database for very conserved regions of protein families (that do not have gaps in the sequence alignment) and then counted the relative frequencies of amino acids and their substitution probabilities.

Then, they calculated a log-odds score for each of the 210 possible substitutions of the 20 standard amino acids. All BLOSUM matrices are based on observed alignments; they are not extrapolated from comparisons of closely related proteins like the PAM Matrices.

Several sets of BLOSUM matrices exist using different alignment databases, named with numbers. BLOSUM matrices with high numbers are designed for comparing closely related sequences, while those with low numbers are designed for comparing distant related sequences. For example, BLOSUM80 is used for less divergent alignments, and BLOSUM45 is used for more divergent alignments.

The matrices were created by merging (clustering) all sequences that were more similar than a given percentage into one single sequence and then comparing those sequences (that were all more divergent than the given percentage value) only; thus reducing the contribution of closely related sequences. The percentage used was appended to the name, giving BLOSUM80 for example where sequences that were more than 80% identical were clustered.

Scores within a BLOSUM are log-odds scores that measure, in an alignment, the logarithm for the ratio of the likelihood of two amino acids appearing with a biological sense and the likelihood of the same amino acids appearing by chance. The matrices are based on the minimum percentage identity of the aligned protein sequence used in calculating them. Every possible identity or substitution is assigned a score based on its observed frequences in the alignment of related proteins. A positive score is given to the more likely substitutions while a negative score is given to the less likely substitutions.

To calculate a BLOSUM matrix, the following equation is used:

$$S_{ij} = \left(\frac{1}{\lambda}\right)\log\left(\frac{p_{ij}}{q_i * q_j}\right)$$

Here, p_{ij} is the probability of two amino acids i and j replacing each other in a homologous sequence, and q_i and q_j are the background probabilities of finding the amino acids and in any protein sequence at random. The factor is a scaling factor, set such that the matrix contains easily computable integer values.

An article in Nature Biotechnology revealed that the BLOSUM62 used for so many years as a standard is not exactly accurate according to the algorithm described by Henikoff and Henikoff. Surprisingly, the miscalculated BLOSUM62 improves search performance.

Brahmagupta Matrix

The following matrix was given by Indian mathematician Brahmagupta:

$$B(x,y)=\begin{bmatrix} x & y \\ \pm ty & \pm x \end{bmatrix}.$$

It satisfies

$$B(x_1,y_1)B(x_2,y_2)=B(x_1x_2 \pm ty_1y_2, x_1y_2 \pm y_1x_2).$$

Powers of the matrix are defined by

$$B^n=\begin{bmatrix} x & y \\ ty & x \end{bmatrix}^n=\begin{bmatrix} x_n & y_n \\ ty_n & x_n \end{bmatrix}\equiv B_n.$$

The x_n and y_n are called Brahmagupta polynomials. The Brahmagupta matrices can be extended to negative integers:

$$B^{-n}=\begin{bmatrix} x & y \\ ty & x \end{bmatrix}^{-n}=\begin{bmatrix} x_{-n} & y_{-n} \\ ty_{-n} & x_{-n} \end{bmatrix}\equiv B_{-n}.$$

Cabibbo–Kobayashi–Maskawa Matrix

In the Standard Model of particle physics, the Cabibbo–Kobayashi–Maskawa matrix (CKM matrix, quark mixing matrix, sometimes also called KM matrix) is a unitary matrix which contains information on the strength of flavour-changing weak decays. Technically, it specifies the mismatch of quantum states of quarks when they propagate freely and when they take part in the weak interactions.

It is important in the understanding of CP violation. A precise mathematical definition of this matrix is given in the article on the formulation of the standard model. This matrix was introduced for three generations of quarks by Makoto Kobayashi and Toshihide Maskawa, adding one generation to the matrix previously introduced by Nicola Cabibbo. This matrix is also an extension of the GIM mechanism, which only includes two of the three current families of quarks.

The Matrix

In 1963, Nicola Cabibbo introduced the Cabibbo angle (θ_c) to preserve the universality of the weak interaction. Cabibbo was inspired by previous work by Murray Gell-Mann and Maurice Lévy, on the effectively rotated nonstrange and strange vector and axial weak currents, which he references.

In light of current knowledge (quarks were not yet theorized), the Cabibbo angle is related to the relative probability that down and strange quarks decay into up quarks ($|V_{uc}|^2$ and $|V_{us}|^2$ respectively). In particle physics parlance, the object that couples to the up quark via charged-current weak interaction is a superposition of down-type quarks, here denoted by $d2$. Mathematically this is:

$$d' = V_{ud}d + V_{us}s,$$

or using the Cabbibo angle:

$$d' = \cos\theta_c d + \sin\theta_c s.$$

Using the currently accepted values for $|V_{ud}|$ and $|V_{us}|$, the Cabbibo angle can be calculated using

$$\tan\theta_c = \frac{|V_{us}|}{|V_{ud}|} = \frac{0.2257}{0.97419} \rightarrow \theta_c = 13.04^\circ.$$

When the charm quark was discovered in 1974, it was noticed that the down and strange quark could decay into either the up or charm quark, leading to two sets of equations:

$$d' = V_{ud}d + V_{us}s;$$

$$s' = V_{cd}d + V_{cs}s,$$

or using the Cabibbo angle:

$$d' = \cos\theta_c d + \sin\theta_c s;$$

$$s' = -\sin\theta_c d + \cos\theta_c s.$$

This can also be written in matrix notation as:

$$\begin{bmatrix} d' \\ s' \end{bmatrix} = \begin{bmatrix} V_{ud} & V_{us} \\ V_{cd} & V_{cs} \end{bmatrix} \begin{bmatrix} d \\ s \end{bmatrix},$$

or using the Cabibb]\o angle

$$\begin{bmatrix} d' \\ s' \end{bmatrix} = \begin{bmatrix} \cos\theta_c & \sin\theta_c \\ -\sin\theta_c & \cos\theta_c \end{bmatrix} \begin{bmatrix} d \\ s \end{bmatrix},$$

where the various $|V_{ij}|^2$ represent the probability that the quark of i flavour decays into a quark of j flavour. This 2×2 rotation matrix is called the Cabibbo matrix. Observing that CP-violation could not be explained in a four-quark model, Kobayashi and Maskawa generalized the Cabbibo matrix into the Cabibbo–Kobayashi–Maskawa matrix (or CKM matrix) to keep track of the weak decays of three generations of quarks:

$$\begin{bmatrix} d' \\ s' \\ b' \end{bmatrix} = \begin{bmatrix} V_{ud} & V_{us} & V_{ub} \\ V_{cd} & V_{cs} & V_{cb} \\ V_{td} & V_{ts} & V_{tb} \end{bmatrix} \begin{bmatrix} d \\ s \\ b \end{bmatrix}.$$

On the left is the weak interaction doublet partners of up-type quarks, and on the right is the CKM matrix along with a vector of mass eigenstates of down-type quarks. The CKM matrix describes the probability of a transition from one quark i to another quark j. These transitions are proportional to $|V_{ij}|^2$.

Currently, the best determination of the magnitudes of the CKM matrix elements is:

$$\begin{bmatrix} |V_{ud}| & |V_{us}| & |V_{ub}| \\ |V_{cd}| & |V_{cs}| & |V_{cb}| \\ |V_{td}| & |V_{ts}| & |V_{tb}| \end{bmatrix} =$$

$$\begin{bmatrix} 0.97428 \pm 0.00015 & 0.2253 \pm 0.0007 & 0.00347^{+0.00016}_{-0.00012} \\ 0.2252 \pm 0.0007 & 0.97345^{+0.00015}_{-0.00016} & 0.0410^{+0.0011}_{-0.0007} \\ 0.00862^{+0.00026}_{-0.00020} & 0.0403^{+0.0011}_{-0.0007} & 0.999152^{+0.000030}_{-0.000045} \end{bmatrix}.$$

Note that the choice of usage of down-type quarks in the definition is purely arbitrary and does not represent some sort of deep physical asymmetry between up-type and down-type quarks.

We could just as easily define the matrix the other way around, describing weak interaction partners of mass eigenstates of up-type quarks, $u2$, $c2$ and $t2$, in terms of u, c, and t. Since the CKM matrix is unitary (and therefore its inverse is the same as its conjugate transpose), we would obtain essentially the same matrix.

Counting

To proceed further, it is necessary to count the number of parameters in this matrix, V which appear in experiments, and therefore are physically important. If there are N generations of quarks ($2N$ flavours) then

- An $N \times N$ unitary matrix (that is, a matrix V such that $VV^{\dagger} = I$, where $V^{\dagger}$ is the conjugate transpose of V and I is the identity matrix) requires N^2 real parameters to be specified.
- $2N - 1$ of these parameters are not physically significant, because one phase can be absorbed into each quark field (both of the mass eigenstates, and of the weak eigenstates), but an overall common phase is unobservable. Hence, the total number of free variables independent of the choice of the phases of basis vectors is $N^2 - (2N - 1) = (N - 1)^2$.
    - o Of these, $N(N - 1)/2$ are rotation angles called quark *mixing angles*.
    - o The remaining $(N - 1)(N - 2)/2$ are complex phases, which cause CP violation.

For the case $N = 2$, there is only one parameter which is a mixing angle between two generations of quarks. Historically, this was the first version of CKM matrix when only two generations were known. It is called the Cabibbo angle after its inventor Nicola Cabibbo.

For the Standard Model case ($N = 3$), there are three mixing angles and one CP-violating complex phase.[7]

Observations and Predictions

Cabibbo's idea originated from a need to explain two observed phenomena:

1. the transitions $u \leftrightarrow d$, $e \leftrightarrow \nu_e$, and $\grave{\imath} \leftrightarrow \nu_{\grave{\imath}}$ had similar amplitudes.
2. the transitions with change in strangeness $\Delta S = 1$ had amplitudes equal to 1/4 of those with $\Delta S = 0$.

Cabibbo's solution consisted of postulating weak universality to resolve the first issue, along with a mixing angle θ_c, now called the *Cabibbo angle*, between the *d* and *s* quarks to resolve the second.

For two generations of quarks, there are no CP violating phases, as shown by the counting of the previous section. Since CP violations were seen in neutral kaon decays already in 1964, the emergence of the Standard Model soon after was a clear signal of the existence of a third generation of quarks, as pointed out in 1973 by Kobayashi and Maskawa. The discovery of the bottom quark at Fermilab (by Leon Lederman's group) in 1976 therefore immediately started off the search for the missing third-generation quark, the top quark.

Weak Universality

The constraints of unitarity of the CKM-matrix on the diagonal terms can be written as

$$\sum_k |V_{ik}|^2 = \sum_i |V_{ik}|^2 = 1$$

for all generations *i*. This implies that the sum of all couplings of any of the up-type quarks to all the down-type quarks is the same for all generations. This relation is called *weak universality* and was first pointed out by Nicola Cabibbo in 1967. Theoretically it is a consequence of the fact that all SU(2) doublets couple with the same strength to the vector bosons of weak interactions. It has been subjected to continuing experimental tests.

The Unitarity Triangles

The remaining constraints of unitarity of the CKM-matrix can be written in the form

$$\sum_k V_{ik} V_{jk}^* = 0.$$

For any fixed and different i and j, this is a constraint on three complex numbers, one for each k, which says that these numbers form the sides of a triangle in the complex plane. There are six choices of i and j (three independent), and hence six such triangles, each of which is called a *unitary triangle.* Their shapes can be very different, but they all have the same area, which can be related to the CP violating phase. The area vanishes for the specific parameters in the Standard Model for which there would be no CP violation. The orientation of the triangles depend on the phases of the quark fields.

Since the three sides of the triangles are open to direct experiment, as are the three angles, a class of tests of the Standard Model is to check that the triangle closes. This is the purpose of a modern series of experiments under way at the Japanese BELLE and the American BaBar experiments, as well as at LHCb in CERN, Switzerland.

Parameterizations

Four independent parameters are required to fully define the CKM matrix. Many parameterizations have been proposed, and three of the most common ones are shown below.

KM Parameters

The original parameterization of Kobayashi and Maskawa used three angles (θ_1, θ_2, θ_3) and a CP-violating phase (δ). Cosines and sines of the angles are denoted c_i and s_i, respectively. θ_1 is the Cabibbo angle.

$$\begin{bmatrix} c_1 & -s_1c_3 & -s_1s_3 \\ s_1c_2 & c_1c_2c_3 - s_2s_3e^{i\delta} & c_1c_2s_3 + s_2c_3e^{i\delta} \\ s_1s_2 & c_1s_2c_3 + c_2s_3e^{i\delta} & c_1s_2s_3 - c_2c_3e^{i\delta} \end{bmatrix}.$$

"Standard" Parameters

A "standard" parameterization of the CKM matrix uses three Euler angles (θ_{12}, θ_{23}, θ_{13}) and one CP-violating phase (δ_{13}). Couplings between quark generation i and j vanish if $\theta_{ij} = 0$. Cosines and sines of the angles are denoted c_{ij} and s_{ij}, respectively. θ_{12} is the Cabibbo angle.

$$\begin{bmatrix} 1 & 0 & 0 \\ 0 & c_{23} & s_{23} \\ 0 & -s_{23} & c_{23} \end{bmatrix} \begin{bmatrix} c_{13} & 0 & s_{13}e^{-i\delta_{13}} \\ 0 & 1 & 0 \\ -s_{13}e^{i\delta_{13}} & 0 & c_{13} \end{bmatrix} \begin{bmatrix} c_{12} & s_{12} & 0 \\ -s_{12} & c_{12} & 0 \\ 0 & 0 & 1 \end{bmatrix}$$

$$= \begin{bmatrix} c_{12}c_{13} & s_{12}c_{13} & s_{13}e^{-i\delta_{13}} \\ -s_{12}c_{23} - c_{12}s_{23}s_{13}e^{i\delta_{13}} & c_{12}c_{23} - s_{12}s_{23}s_{13}e^{i\delta_{13}} & s_{23}c_{13} \\ s_{12}s_{23} - c_{12}c_{23}s_{13}e^{i\delta_{13}} & -c_{12}s_{23} - s_{12}c_{23}s_{13}e^{i\delta_{13}} & c_{23}c_{13} \end{bmatrix}.$$

The currently best known values for the standard parameters are:

$\theta_{12} = 13.04\pm0.05°$, $\theta_{13} = 0.201\pm0.011°$, $\theta_{23} = 2.38\pm0.06°$, and $\delta_{13} = 1.20\pm0.08$.

Wolfenstein Parameters

A third parameterization of the CKM matrix was introduced by Lincoln Wolfenstein with the four parameters λ, A, ρ, and η.

The four Wolfenstein parameters have the property that all are of order 1 and are related to the "standard" parameterization:

$$\lambda = s_{12}$$
$$A\lambda^2 = s_{23}$$
$$A\lambda^3(\rho - i\eta) = s_{13}\geq^{i\delta}$$

The Wolfenstein parameterization of the CKM matrix, to order λ^3, is

$$\begin{bmatrix} 1-\lambda^2/2 & \lambda & A\lambda^3(\rho - i\eta) \\ -\lambda & 1-\lambda^2/2 & A\lambda^2 \\ A\lambda^3(1-\rho-i\eta) & -A\lambda^2 & 1 \end{bmatrix}.$$

The CP violation can be determined by measuring $\rho - i\eta$.

Using the values of the previous section for the CKM matrix, the best determination of the Wolfenstein parameters is:

$\lambda = 0.2257+0.0009$

$- 0.0010$, $A = 0.814+0.021$

$- 0.022$, $\rho = 0.135+0.031$

$- 0.016$, and $\eta = 0.349+0.015$

$- 0.017$.

Nobel Prize

In 2008, Kobayashi and Maskawa shared one half of the Nobel Prize in Physics "for the discovery of the origin of the broken symmetry which predicts the existence of at least three families of quarks in nature".

Some physicists were reported to harbor bitter feelings about the fact that the Nobel Prize committee failed to reward the work of Cabibbo, whose prior work was closely related to that of Kobayashi and Maskawa.

Asked for a reaction on the prize, Cabibbo preferred to give no comment.

Chapter 12

Metzler Matrix

In mathematics, a Metzler matrix is a matrix in which all the off-diagonal components are nonnegative (equal to or greater than zero)

$$\forall_{i \neq j} x_{ij} \geq 0.$$

It is named after the American economist Lloyd Metzler. Metzler matrices appear in stability analysis of time delayed differential equations and positive linear dynamical systems. Their properties can be derived by applying the properties of nonnegative matrices to matrices of the form $M + aI$ where M is a Metzler matrix.

Matrix Representation

Matrix representation is a method used by a computer language to store matrices of more than one dimension in memory.

Fortran and C use different schemes. Fortran uses "Column Major", in which all the elements for a given column are stored contiguously in memory. C uses "Row Major", which stores all the elements for a given row contiguously in memory.

LAPACK defines various matrix representations in memory. There is also Sparse matrix representation and Morton-order matrix representation.

According to the documentation, in LAPACK the unitary matrix representation is optimized. Some languages such as Java store matrices using Iliffe vectors. These are particularly useful for storing irregular matrices. Matrices are of primary importance in linear algebra.

Basic Mathematical Operations

An m × n (read as m by n) order matrix is a set of numbers arranged in m rows and n columns. Matrices of the same order can be added by adding the corresponding elements.

Two matrices can be multiplied, the condition being that the number of columns of the first matrix is equal to the number of rows of the second matrix. Hence, if an m × n matrix is multiplied with an n × r matrix, then the resultant matrix will be of the order m × r.

Operations like row operations or column operations can be performed on a matrix, using which we can obtain the inverse of a matrix. The inverse may be obtained by determining the adjoint as well.

Basics of 2D Array

The mathematical definition of a matrix finds applications in computing and database management, a basic starting point being the concept of arrays. A two-dimensional array can function exactly like a matrix. Two-dimensional arrays can be visualized as a table consisting of rows and columns.

- Int a, declares an integer array of 3 rows and 4 columns. Index of row will start from 0 and will go up to 2.
- Similarly, index of column will start from 0 and will go up to 3.

	Column 0	*Column 1*	*Column 2*	*Column 3*
row 0	a[0][0]	a[0][1]	a[0][2]	[0][3]
row 1	a[1][0]	a[1][1]	a[1][2]	[1][3]
row 2	a[2][0]	a[2][1]	a[2][2]	[2][3]

This table shows arrangement of elements with their indices.

Initializing Two-Dimensional arrays: Two-Dimensional arrays may be initialized by providing a list of initial values.

int a[2][3] = {1,2,3,4,5,6,} or int a[2][3] = {{2,3,4},{4,4.5}};

Calculation of Address : An m x n matrix (a[1...m][1...n]) where the row index varies from 1 to m and column index from 1 to n,a_{ij} denotes the number in the ith row and the jth column.

In the computer memory, all elements are stored linearly using contiguous addresses. Therefore, in order to store a two-dimensional matrix a, two dimensional address space must be mapped to one dimensional address space. In the computer's memory matrices are stored in either Row-major order or Column-major order form.

Modal Matrix

In linear algebra, the modal matrix is used in the diagonalization process involving eigenvalues and eigenvectors.

Assume a linear system of the following form:

$$\frac{d}{dt}X = A^{*}X + B^{*}U$$

where X is $n\times1$, A is $n\times n$, and B is $n\times1$. X typically represents the state vector, and U the system input. Specifically the modal matrix M is the $n\times n$ matrix formed with the eigenvectors of A as columns in M. It is utilized in

$$M^{-1}AM = D$$

where D is an $n\times n$ diagonal matrix with the eigenvalues of A on the main diagonal of D and zeros elsewhere. This process is also known as the similarity transform.

Moore Matrix

In linear algebra, a Moore matrix, introduced by E. H. Moore (1896), is a matrix defined over a finite field.

When it is a square matrix its determinant is called a Moore determinant (this is unrelated to the Moore determinant of a quaternionic Hermitian matrix). The Moore matrix has successive powers of the Frobenius automorphism applied to the first column, so it is an $m \times n$ matrix

$$M = \begin{bmatrix} \alpha_1 & \alpha_1^{q} & \dots & \alpha_1^{q^{n-1}} \\ \alpha_2 & \alpha_2^{q} & \dots & \alpha_2^{q^{n-1}} \\ \alpha_3 & \alpha_3^{q} & \dots & \alpha_3^{q^{n-1}} \\ \vdots & \vdots & \ddots & \vdots \\ \alpha_m & \alpha_m^{q} & \dots & \alpha_m^{q^{n-1}} \end{bmatrix}$$

or

$$M_{i,j} = \alpha_i^{q^{j-1}}$$

for all indices i and j. (Some authors use the transpose of the above matrix.)

The Moore determinant of a square Moore matrix (so $m = n$) can be expressed as:

$$\det(V) = \prod_{\mathbf{c}} \left(c_1\alpha_1 + \cdots + c_n\alpha_n\right),$$

where c runs over a complete set of direction vectors, made specific by having the last non-zero entry equal to 1, i.e.

$$\det(V) = \prod_{1 \le i \le n} \prod_{c_1, \ldots, c_{i-1}} \left(c_1\alpha_1 + \cdots + c_{i-1}\alpha_{i-1} + \alpha_i\right).$$

In particular the Moore determinant vanishes if and only if the elements in the left hand column are linearly independent over the finite field of order q.

So it is analogous to the Wronskian of several functions. Dickson used the Moore determinant in finding the modular invariants of the general linear group over a finite field.

Orthostochastic Matrix

In mathematics, an orthostochastic matrix is a doubly stochastic matrix whose entries are the square of the absolute value of some orthogonal matrix.

The detailed definition is as follows. An square matrix B of size n is doubly stochastic (or *bistochastic*) if all its rows and columns sum to 1 and all its entries are nonnegative real numbers, each of whose rows and columns sums to 1.

It is orthostochastic if there exists an orthogonal matrix O such that

$$B_{ij} = O_{ij}^2 \text{ for } i, j = 1, \ldots, n.$$

All 2-by-2 doubly stochastic matrices are orthostochastic (and also unistochastic) since for any

$$B = \begin{bmatrix} a & 1-a \\ 1-a & a \end{bmatrix}$$

we find the corresponding orthogonal matrix

$$O = \begin{bmatrix} \cos\phi & \sin\phi \\ -\sin\phi & \cos\phi \end{bmatrix},$$

with $\cos^2\phi = a$, such that $B_{ij} = O_{ij}^2$.

For larger n the sets of bistochastic matrices includes the set of unistochastic matrices, which includes the set of orthostochastic matrices and these inclusion relations are proper.

Overlap Matrix

The overlap matrix is a square matrix, used in quantum chemistry to describe the inter-relationship of a set of basis vectors of a quantum system. In particular, if the vectors are orthogonal to one another, the overlap matrix will be diagonal. In addition, if the basis vectors form an orthonormal set, the overlap matrix will be the identity matrix. The overlap matrix is always $n \times n$, where n is the number of basis functions used. It is a kind of Gramian matrix.

In general, the overlap matrix is defined as:

$$\mathbf{S}_{jk} = \langle b_j \mid b_k \rangle = \int \Psi_j^* \Psi_k d\tau$$

where

$|b_j\rangle$ is the j-th basis ket (vector), and

Ψ_j is the j-th wavefunction, defined as :

$$\Psi_j(x) = \langle x \mid b_j \rangle.$$

In particular, if the set is normalized (though not necessarily orthogonal) then the diagonal elements will be identically 1 and the magnitude of the off-diagonal elements less than or equal to one with equality if and only if there is linear dependence in the basis set as per the Cauchy–Schwarz inequality.

Moreover, the matrix is always positive definite; that is to say, the eigenvalues are all strictly positive.

Packed Storage Matrix

A packed storage matrix, also known as packed matrix, is a term used in programming for representing an $m \times n$ matrix.

It is a more compact way than an m-by-n rectangular array by exploiting a special structure of the matrix.

Typical examples of matrices that can take advantage of packed storage include:

- symmetric or hermitian matrix
- Triangular matrix
- Banded matrix.

Code Examples (Fortran)

Both of the following storage schemes are used extensively in BLAS and LAPACK.

An example of packed storage for hermitian matrix:

```
complex:: A(n,n) ! a hermitian matrix
complex:: AP(n*(n+1)/2) ! packed storage for A
! the lower triangle of A is stored column-by-column in AP.
! unpacking the matrix AP to A
do j=1,n
  k = j*(j-1)/2
  A(1:j,j) = AP(1+k:j+k)
  A(j,1:j-1) = conjg(AP(1+k:j-1+k))
end do
```

An example of packed storage for banded matrix:

```
real:: A(m,n) ! a banded matrix with kl subdiagonals and ku superdiagonals
real:: AP(-kl:ku,n) ! packed storage for A
! the band of A is stored column-by-column in AP. Some elements of AP are
unused.
! unpacking the matrix AP to A
do j=1,n
  forall(i=max(1,j-kl):min(m,j+ku)) A(i,j) = AP(i-j,j)
end do
print *,AP(0,:) ! the diagonal
```

Persymmetric Matrix

In mathematics, persymmetric matrix may refer to:

1. A square matrix which is symmetric in the northeast-to-southwest diagonal; or
2. A square matrix such that the values on each line perpendicular to the main diagonal are the same for a given line.

The first definition is the most common in the recent literature. The designation "Hankel matrix" is often used for matrices satisfying the property in the second definition.

Definition 1

Let $A = (a_{ic\,j})$ be an $n \times n$ matrix. The first definition of *persymmetric* requires that

$$a_{ij} = a_{n-j+1,n-i+1} \text{for all } i, j.$$

For example, 5-by-5 persymmetric matrices are of the form

$$A = \begin{bmatrix} a_{11} & a_{12} & a_{13} & a_{14} & a_{15} \\ a_{21} & a_{22} & a_{23} & a_{24} & a_{14} \\ a_{31} & a_{32} & a_{33} & a_{23} & a_{13} \\ a_{41} & a_{42} & a_{32} & a_{22} & a_{12} \\ a_{51} & a_{41} & a_{31} & a_{21} & a_{11} \end{bmatrix}.$$

This can be equivalently expressed as $AJ = JA^{\mathrm{T}}$ where J is the exchange matrix.

A symmetric matrix is a matrix whose values are symmetric in the northwest-to-southeast diagonal.

If a symmetric matrix is rotated by 90°, it becomes a persymmetric matrix. Symmetric persymmetric matrices are sometimes called bisymmetric matrices.

Definition 2

The second definition is due to Thomas Muir. It says that the square matrix $A = (a_{ij})$ is persymmetric if a_{ij} depends only on $i + j$.

Persymmetric matrices in this sense, or Hankel matrices as they are often called, are of the form

$$A = \begin{bmatrix} r_1 & r_2 & r_3 & \cdots & r_n \\ r_2 & r_3 & r_4 & \cdots & r_{n+1} \\ r_3 & r_4 & r_5 & \cdots & r_{n+2} \\ \vdots & \vdots & \vdots & \ddots & \vdots \\ r_n & r_{n+1} & r_{n+2} & \cdots & r_{2n-1} \end{bmatrix}.$$

A persymmetric determinant is the determinant of a persymmetric matrix.

A matrix for which the values on each line parallel to the main diagonal are constant, is called a Toeplitz matrix.

Pick Matrix

In mathematics, a Pick matrix, named after Georg Pick, is a matrix which occurs in the study of interpolation problems for analytic functions.

It is defined as

$$P = \left[\frac{u_i^* u_i - v_j^* v_j}{1 - f_i^* f_j} \right]_{i,j=0}^{n-1}$$

where $(u_i)$and (u_i) are column vectors of dimension p and q, respectively; and z^* denotes the complex conjugate of z. Here f_i are complex numbers inside the unit circle. That is,

$$| f_i | < 1.$$

Polynomial Matrix

In mathematics, a polynomial matrix or sometimes matrix polynomial is a matrix whose elements are univariate or multivariate polynomials. A λ-matrix is a matrix whose elements are polynomials in λ.

A univariate polynomial matrix P of degree p is defined as:

$$P = \sum_{n=0}^{p} A(n)x^n = A(0) + A(1)x + A(2)x^2 + \cdots + A(p)x^p$$

where *A(i)* denotes a matrix of constant coefficients, and *A(p)* is non-zero.

Thus a polynomial matrix is the matrix-equivalent of a polynomial, with each element of the matrix satisfying the definition of a polynomial of degree p.

An example 3×3 polynomial matrix, degree 2:

$$P = \begin{pmatrix} 1 & x^2 & x \\ 0 & 2x & 2 \\ 3x+2 & x^2-1 & 0 \end{pmatrix} = \begin{pmatrix} 1 & 0 & 0 \\ 0 & 0 & 2 \\ 2 & -1 & 0 \end{pmatrix} +$$

$$\begin{pmatrix} 0 & 0 & 1 \\ 0 & 2 & 0 \\ 3 & 0 & 0 \end{pmatrix} x + \begin{pmatrix} 0 & 1 & 0 \\ 0 & 0 & 0 \\ 0 & 1 & 0 \end{pmatrix} x^2.$$

We can express this by saying that for a ring R, the rings $M_n(R[X])$ and $(M_n(R))[X]$ are isomorphic.

Properties

- A polynomial matrix over a field with determinant equal to a non-zero element of that field is called unimodular, and has an inverse that is also a polynomial matrix. Note that the only scalar unimodular polynomials are polynomials of degree 0 - nonzero constants, because an inverse of an arbitrary polynomial of higher degree is a rational function.
- The roots of a polynomial matrix over the complex numbers are the points in the complex plane where the matrix loses rank.

Note that polynomial matrices are *not* to be confused with monomial matrices, which are simply matrices with exactly one non-zero entry in each row and column.

If by λ we denote any element of the field over which we constructed the matrix, by I the identity matrix, and we let A be a polynomial matrix, then the matrix $\lambda I\text{-}A$ is the characteristic matrix of the matrix A. Its determinant, $|\lambda I\text{-}A|$ is the characteristic polynomial of the matrix A.

R-matrix

The term R-matrix has several meanings, depending on the field of study.

The term R-matrix is used in connection with the Yang–Baxter equation. This is an equation which was first introduced in the field of statistical mechanics, taking its name from independent work of C. N. Yang and R. J. Baxter. The classical R-matrix arises in the definition of the classical Yang–Baxter equation.

In quasitriangular Hopf algebra, where the R-matrix is a solution of the Yang–Baxter equation. The numerical modeling of diffraction gratings in optical science can be performed using the R-matrix propagation algorithm..

There is a method in computational quantum mechanics for studying scattering known as the R-matrix.

This method was originally formulated for studying resonances in nuclear scattering by Wigner and Eisenbud. Using that work as a basis, an R-matrix method was developed for electron, positron and photon scattering by atoms. This approach was later adapted for electron, positron and photon scattering by molecules.

Regular Hadamard Matrix

In mathematics a regular Hadamard matrix is a Hadamard matrix whose row and column sums are all equal. While the order of a Hadamard matrix must be 1, 2, or a multiple of 4, regular Hadamard matrices carry the further restriction that the order be a perfect square. The excess, denoted $E(H)$, of a Hadamard matrix H of order n is defined to be the sum of the entries of H. The excess satisfies the bound $|E(H)| \leq n^{3/2}$. A Hadamard matrix attains this bound if and only if it is regular.

If $n = 4u^2$ is the order of a regular Hadamard matrix, then the excess is $\pm 8u^3$ and the row and column sums all equal $\pm 2u$. It follows that each row has $2u^2 \pm u$ positive entries and $2u^2$ – " u negative entries. The orthogonality of rows implies that any two distinct rows have exactly $u^2 \pm u$ positive entries in common. If H is interpreted as the incidence matrix of a block design, with 1 representing incidence and –1 representing non-incidence, then H corresponds to a 2-(v,k,λ) design with parameters $(4u^2, 2u^2 \pm u, u^2 \pm u)$. A design with these parameters is called a Menon design.

A number of methods for constructing regular Hadamard matrices are known, and some exhaustive computer searches have been done for regular Hadamard matrices with specified symmetry groups, but it is not known whether every even perfect square is the order of a regular Hadamard matrix. Bush-type Hadamard matrices are regular Hadamard matrices of a special form, and are connected with finite projective planes.

Rotation Matrix

In linear algebra, a rotation matrix is a matrix that is used to perform a rotation in Euclidean space. For example the matrix

$$R = \begin{bmatrix} \cos\theta & -\sin\theta \\ \sin\theta & \cos\theta \end{bmatrix}$$

rotates points in the xy-Cartesian plane counterclockwise through an angle θ about the origin of the Cartesian coordinate system. To perform the rotation using a rotation matrix R, the position of each point must be represented by a column vector v, containing the coordinates of the point. A rotated vector is obtained by using the matrix multiplication Rv. Since matrix multiplication has no effect on the zero vector (i.e., on the coordinates of the origin), rotation matrices can only be used to describe rotations about the origin of the coordinate system.

Rotation matrices provide a simple algebraic description of such rotations, and are used extensively for computations in geometry, physics, and computer graphics. In two-dimensional space, a rotation can be simply described by an angle θ of rotation, but it can be also represented by the four entries of a rotation matrix with two rows and two columns.

In three-dimensional space, every rotation can be interpreted as a rotation by a given angle about a single fixed axis of rotation, and hence it can be simply described by an angle and a vector with three entries. However, it can also be represented by the nine entries of a rotation matrix with three rows and three columns. The notion of rotation is not commonly used in dimensions higher than 3; there is a notion of a rotational displacement, which can be represented by a matrix, but not associated single axis or angle.

Rotation matrices are square matrices, with real entries. More specifically they can be characterized as orthogonal matrices with determinant 1:

$$R^T = R^{-1}, \det R = 1$$

The set of all such matrices of size n forms a (generally not commutative) group, known as the special orthogonal group SO(n).

In some literature, the term *rotation* is generalized to include improper rotations, characterized by orthogonal matrices with determinant -1 (instead of +1). These combine proper rotations with *reflections* (which invert orientation). In this sense the members of the (general) orthogonal group O(n) may also be called rotation matrices. In other cases, where reflections are not being considered, the label *proper* may be dropped unless demands of clarity make it wise to specify it. This is the reason it is more frequent to associate the term *rotation matrix* with members of SO(n) (as in the rest of this article) while it may also be found referring to members of O(n).

Common Rotations

Particularly useful are the matrices for 90° and 180° rotations:

$$R(90^{\circ}) = \begin{bmatrix} 0 & -1 \\ [3pt]1 & 0 \end{bmatrix}$$ (90° counterclockwise rotation)

$$R(180^{\circ}) = \begin{bmatrix} -1 & 0 \\ [3pt]0 & -1 \end{bmatrix}$$ (180° rotation in either direction – a half-turn)

$$R(270^{\circ}) = \begin{bmatrix} 0 & 1 \\ [3pt]-1 & 0 \end{bmatrix}$$ (270° counterclockwise rotation, the same as a 90° clockwise rotation)

In Three Dimensions

Basic Rotations

The following three basic (gimbal-like) rotation matrices rotate vectors about the x, y, or z axis, in three dimensions:

$$\begin{aligned} R_x(\theta) &= \begin{bmatrix} 1 & 0 & 0 \\ 0 & \cos\theta & -\sin\theta \\ [3pt]0 & \sin\theta & \cos\theta \\ [3pt] \end{bmatrix} \\ [6pt]R_y(\theta) &= \begin{bmatrix} \cos\theta & 0 & \sin\theta \\ [3pt]0 & 1 & 0 \\ [3pt]-\sin\theta & 0 & \cos\theta \end{bmatrix} \\ [6pt]R_z(\theta) &= \begin{bmatrix} \cos\theta & -\sin\theta & 0 \\ [3pt]\sin\theta & \cos\theta & 0 \\ [3pt]0 & 0 & 1 \end{bmatrix}. \end{aligned}$$

Each of these basic vector rotations typically appears counter-clockwise when the axis about which they occur points towards the observer, and the coordinate system is right-handed. R_z, for instance, would rotate towards the y-axis a vector aligned with the x-axis. This is similar to the rotation produced by the above mentioned 2-D rotation matrix.

General Rotations

Other rotation matrices can be obtained from these three using matrix multiplication. For example, the product

$$R_x(\gamma)R_y(\beta)R_z(\alpha)$$

represents a rotation whose yaw, pitch, and roll are α, β, and γ, respectively. Similarly, the product

$$R_z(\gamma)R_x(\beta)R_y(\alpha)$$

represents a rotation whose Euler angles are α, β, and γ (using the y-x-z convention for Euler angles). The orthogonal matrix (post-multiplying a column vector) corresponding to a clockwise/left-handed rotation with Euler angles ϕ,θ,ψ, with x-y-z convention, is given by:

$$R_z(\psi)R_y(\theta)R_x(\phi)=$$

$$\begin{bmatrix} \cos\theta\cos\psi & -\cos\phi\sin\psi+\sin\phi\sin\theta\cos\psi & \sin\phi\sin\psi+\cos\phi\sin\theta\cos\psi \\ \cos\theta\sin\psi & \cos\phi\cos\psi+\sin\phi\sin\theta\sin\psi & -\sin\phi\cos\psi+\cos\phi\sin\theta\sin\psi \\ -\sin\theta & \sin\phi\cos\theta & \cos\phi\cos\theta \end{bmatrix}$$

Geometry

In Euclidean geometry, a rotation is an example of an isometry, a transformation that moves points without changing the distances between them.

Rotations are distinguished from other isometries by two additional properties: they leave (at least) one point fixed, and they leave "handedness" unchanged. By contrast, a translation moves every point, a reflection exchanges left- and right-handed ordering, and a glide reflection does both.

A rotation that does not leave "handedness" unchanged is an improper rotation or a rotoinversion. If we take the fixed point as the origin of a Cartesian coordinate system, then every point can be given coordinates as a displacement from the origin. Thus we may work with the vector space of displacements instead of the points themselves. Now suppose $(p_1,\ldots,p_n)$ are the coordinates of the vector p from the origin, O, to point P. Choose an orthonormal basis for our coordinates; then the squared distance to P, by Pythagoras, is

$$d^2(O,P) = \|\mathbf{p}\|^2 = \sum_{r=1}^{n} p_r^2$$

which we can compute using the matrix multiplication

$$\|\mathbf{p}\|^2 = \begin{bmatrix} p_1 \cdots p_n \end{bmatrix} \begin{bmatrix} p_1 \\ \vdots \\ p_n \end{bmatrix} = \mathbf{p}^T\mathbf{p}.$$

A geometric rotation transforms lines to lines, and preserves ratios of distances between points. From these properties we can show that a rotation is a linear transformation of the vectors, and thus can be written in matrix form, Qp. The fact that a rotation preserves, not just ratios, but distances themselves, we can state as

$$\mathbf{p}^T\mathbf{p} = (Q\mathbf{p})^T(Q\mathbf{p}),$$

or

$$\begin{aligned} \mathbf{p}^T I\mathbf{p} &= (\mathbf{p}^T Q^T)(Q\mathbf{p}) \\ &= \mathbf{p}^T (Q^T Q)\mathbf{p}. \end{aligned}$$

Because this equation holds for all vectors, p, we conclude that every rotation matrix, Q, satisfies the *orthogonality* condition,

$$Q^T Q = I.$$

Rotations preserve handedness because they cannot change the ordering of the axes, which implies the *special matrix* condition,

$$\det Q = +1.$$

Equally important, we can show that any matrix satisfying these two conditions acts as a rotation.

Multiplication

The inverse of a rotation matrix is its transpose, which is also a rotation matrix:

$$\begin{aligned}(Q^T)^T(Q^T) &= QQ^T = I \\ \det Q^T &= \det Q = +1.\end{aligned}$$

The product of two rotation matrices is a rotation matrix:

$$\begin{aligned}(Q_1Q_2)^T(Q_1Q_2) &= Q_2^T(Q_1^TQ_1)Q_2 = I \\ \det(Q_1Q_2) &= (\det Q_1)(\det Q_2) = +1.\end{aligned}$$

For n greater than 2, multiplication of $n\times n$ rotation matrices is not commutative.

$$Q_1 = \begin{bmatrix} 0 & -1 & 0 \\ 1 & 0 & 0 \\ 0 & 0 & 1 \end{bmatrix} \quad Q_2 = \begin{bmatrix} 0 & 0 & 1 \\ 0 & 1 & 0 \\ -1 & 0 & 0 \end{bmatrix}$$

$$Q_1Q_2 = \begin{bmatrix} 0 & -1 & 0 \\ 0 & 0 & 1 \\ -1 & 0 & 0 \end{bmatrix} \quad Q_2Q_1 = \begin{bmatrix} 0 & 0 & 1 \\ 1 & 0 & 0 \\ 0 & 1 & 0 \end{bmatrix}.$$

Noting that any identity matrix is a rotation matrix, and that matrix multiplication is associative, we may summarize all these properties by saying that the $n\times n$ rotation matrices form a group, which for $n > 2$ is non-abelian. Called a special orthogonal group, and denoted by SO(n), SO(n,R), SO_n, or SO_n(R), the group of $n\times n$ rotation matrices is isomorphic to the group of rotations in an n-dimensional space.

This means that multiplication of rotation matrices corresponds to composition of rotations, applied in left-to-right order of their corresponding matrices.

Ambiguities

The interpretation of a rotation matrix can be subject to many ambiguities.

Alias or alibi transformation: The change in a vector's coordinates can be due to a turn of the coordinate system (alias) or a turn of the vector (alibi). Any rotation can be legitimately described both ways, as vectors and coordinate systems actually rotate with respect to each other. Throughout this article, we chose the alibi approach to describe rotations.

Pre-multiplication or post-multiplication: The vector can be pre-multiplied by a rotation matrix (Rv, where v is a column vector), or post-multiplied by it (vR, where v is a row vector). Throughout this article, we described rotations produced by means of a pre-multiplication.

Right- or left-handed coordinates: The matrix and the vector can be represented with respect to a right-handed or left-handed coordinate system. Throughout the article, we assumed a right-handed orientation, unless otherwise specified.

Vectors or forms: The vector space has a dual space of linear forms, and the matrix can act on either vectors or forms.

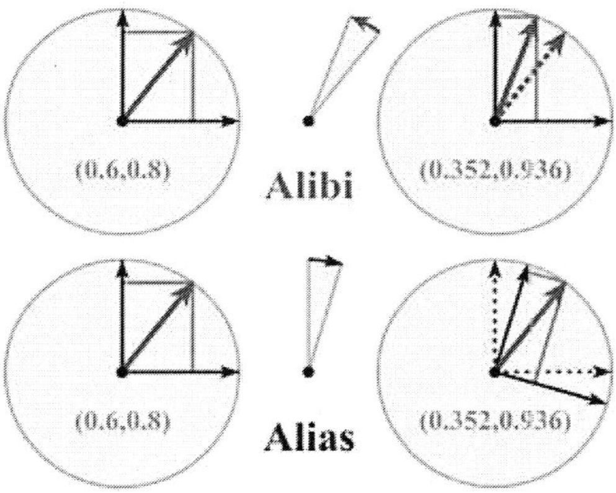

Figure: *Alias and alibi rotations.*

In most cases the effect of the ambiguity is equivalent to the effect of a transposition of the rotation matrix.

Decompositions

Independent Planes

Consider the 3×3 rotation matrix

$$Q = \begin{bmatrix} 0.36 & 0.48 & -0.8 \\ -0.8 & 0.60 & 0 \\ 0.48 & 0.64 & 0.60 \end{bmatrix}.$$

If Q acts in a certain direction, v, purely as a scaling by a factor λ, then we have

$$Q\mathbf{v} = \lambda\mathbf{v},$$

so that

$$\mathbf{0} = (\lambda I - Q)\mathbf{v}.$$

Thus λ is a root of the characteristic polynomial for Q,

$$\begin{aligned} 0 &= \det(\lambda I - Q) \\ &= \lambda^3 - \tfrac{39}{25}\lambda^2 + \tfrac{39}{25}\lambda - 1 \\ &= (\lambda - 1)(\lambda^2 - \tfrac{14}{25}\lambda + 1). \end{aligned}$$

Two features are noteworthy. First, one of the roots (or eigenvalues) is 1, which tells us that some direction is unaffected by the matrix. For rotations in three dimensions, this is the *axis* of the rotation (a concept that has no meaning in any other dimension).

Second, the other two roots are a pair of complex conjugates, whose product is 1 (the constant term of the quadratic), and whose sum is 2 cos θ (the negated linear term). This factorization is of interest for 3×3 rotation matrices because the same thing occurs for all of them. (As special cases, for a null rotation the "complex conjugates" are both 1, and for a 180° rotation they are both −1.)

Furthermore, a similar factorization holds for any $n{\times}n$ rotation matrix. If the dimension, n, is odd, there will be a "dangling" eigenvalue of 1; and for any dimension the rest of the polynomial factors into quadratic terms like the one here (with the two special cases noted). We are guaranteed that the characteristic polynomial will have degree n and thus n eigenvalues.

And since a rotation matrix commutes with its transpose, it is a normal matrix, so can be diagonalized. We conclude that every rotation matrix, when expressed in a suitable coordinate system, partitions into independent rotations of two-dimensional subspaces, at most $^{n}\mathrm{D}_{2}$ of them.

The sum of the entries on the main diagonal of a matrix is called the trace; it does not change if we reorient the coordinate system, and always equals the sum of the eigenvalues.

This has the convenient implication for 2×2 and 3×3 rotation matrices that the trace reveals the angle of rotation, θ, in the two-dimensional (sub-)space.

For a 2×2 matrix the trace is 2 cos(θ), and for a 3×3 matrix it is 1+2 cos(θ). In the three-dimensional case, the subspace consists of all vectors perpendicular to the rotation axis (the invariant direction, with eigenvalue 1). Thus we can extract from any 3×3 rotation matrix a rotation axis and an angle, and these completely determine the rotation.

Sequential Angles

The constraints on a 2×2 rotation matrix imply that it must have the form

$$Q = \begin{bmatrix} a & -b \\ b & a \end{bmatrix}$$

with $a^2+b^2 = 1$. Therefore we may set $a = \cos\theta$ and $b = \sin\theta$, for some angle θ.

To solve for θ it is not enough to look at a alone or b alone; we must consider both together to place the angle in the correct quadrant, using a two-argument arctangent function.

Now consider the first column of a 3×3 rotation matrix,

$$\begin{bmatrix} a \\ b \\ c \end{bmatrix}.$$

Although a^2+b^2 will probably not equal 1, but some value $r^2 < 1$, we can use a slight variation of the previous computation to find a so-called Givens rotation that transforms the column to

$$\begin{bmatrix} r \\ 0 \\ c \end{bmatrix},$$

zeroing b. This acts on the subspace spanned by the x and y axes. We can then repeat the process for the xz subspace to zero c. Acting on the full matrix, these two rotations produce the schematic form

$$Q_{xz}Q_{xy}Q = \begin{bmatrix} 1 & 0 & 0 \\ 0 & * & * \\ 0 & * & * \end{bmatrix}.$$

Shifting attention to the second column, a Givens rotation of the yz subspace can now zero the z value. This brings the full matrix to the form

$$Q_{yz}Q_{xz}Q_{xy}Q = \begin{bmatrix} 1 & 0 & 0 \\ 0 & 1 & 0 \\ 0 & 0 & 1 \end{bmatrix},$$

which is an identity matrix. Thus we have decomposed Q as

$$Q = Q_{xy}^{-1}Q_{xz}^{-1}Q_{yz}^{-1}.$$

An $n \times n$ rotation matrix will have $(n-1)+(n-2)+$ï–$+2+1$, or

$$\sum_{k=1}^{n-1} k = \frac{n(n-1)}{2}$$

entries below the diagonal to zero. We can zero them by extending the same idea of stepping through the columns with a series of rotations in a fixed sequence of planes. We conclude that the set of $n \times n$ rotation matrices, each of which has n^2 entries, can be parameterized by $n(n-1)/2$ angles.

xzx_w	xzy_w	xyx_w	xyz_w
yxy_w	yxz_w	yzy_w	yzx_w
zyz_w	zyx_w	zxz_w	zxy_w
xzx_b	yzx_b	xyx_b	zyx_b
yxy_b	zxy_b	yzy_b	xzy_b
zyz_b	xyz_b	zxz_b	yxz_b

In three dimensions this restates in matrix form an observation made by Euler, so mathematicians call the ordered sequence of three angles Euler angles. However, the situation is somewhat more complicated than we have so far indicated. Despite the small dimension, we actually have considerable freedom in the sequence of axis pairs we use; and we also have some freedom in the choice of angles.

Thus we find many different conventions employed when three-dimensional rotations are parameterized for physics, or medicine, or chemistry, or other disciplines. When we include the option of world axes or body axes, 24 different sequences are possible. And while some disciplines call any sequence Euler angles, others give different names (Euler, Cardano, Tait-Bryan, Roll-pitch-yaw) to different sequences. One reason for the large number of options is that, as noted previously, rotations in three dimensions (and higher) do not commute. If we reverse a given sequence of rotations, we get a different outcome. This also implies that we cannot compose two rotations by adding their corresponding angles. Thus *Euler angles are not vectors*, despite a similarity in appearance as a triple of numbers.

Nested Dimensions

A 3×3 rotation matrix like

$$Q_{3\times 3} = \begin{bmatrix} \cos\theta & -\sin\theta & 0 \\ \sin\theta & \cos\theta & 0 \\ 0 & 0 & 1 \end{bmatrix}$$

suggests a 2×2 rotation matrix,

$$Q_{2\times 2} = \begin{bmatrix} \cos\theta & -\sin\theta \\ \sin\theta & \cos\theta \end{bmatrix},$$

is embedded in the upper left corner:

$$Q_{3\times 3} = \begin{bmatrix} Q_{2\times 2} & \mathbf{0} \\ \mathbf{0}^T & 1 \end{bmatrix}.$$

This is no illusion; not just one, but many, copies of n-dimensional rotations are found within $(n+1)$-dimensional rotations, as subgroups. Each embedding leaves one direction fixed, which in the case of 3×3 matrices is the rotation axis. For example, we have

$$Q_x(\theta) = \begin{bmatrix} 1 & 0 & 0 \\ 0 & \cos\theta & -\sin\theta \\ 0 & \sin\theta & \cos\theta \end{bmatrix},$$

$$Q_y(\theta) = \begin{bmatrix} \cos\theta & 0 & \sin\theta \\ 0 & 1 & 0 \\ -\sin\theta & 0 & \cos\theta \end{bmatrix},$$

$$Q_z(\theta) = \begin{bmatrix} \cos\theta & -\sin\theta & 0 \\ \sin\theta & \cos\theta & 0 \\ 0 & 0 & 1 \end{bmatrix},$$

fixing the x axis, the y axis, and the z axis, respectively. The rotation axis need not be a coordinate axis; if u = (x,y,z) is a unit vector in the desired direction, then

$$
\begin{aligned}
Q_{\mathbf{u}}(\theta) &= \begin{bmatrix} 0 & -z & y \\ z & 0 & -x \\ -y & x & 0 \end{bmatrix} \sin\theta + (I - \mathbf{u}\mathbf{u}^T)\cos\theta + \mathbf{u}\mathbf{u}^T \\
&= \begin{bmatrix} (1-x^2)c_\theta + x^2 & -zs_\theta - xyc_\theta + xy & ys_\theta - xzc_\theta + xz \\ zs_\theta - xyc_\theta + xy & (1-y^2)c_\theta + y^2 & -xs_\theta - yzc_\theta + yz \\ -ys_\theta - xzc_\theta + xz & xs_\theta - yzc_\theta + yz & (1-z^2)c_\theta + z^2 \end{bmatrix} \\
&= \begin{bmatrix} x^2(1-c_\theta) + c_\theta & xy(1-c_\theta) - zs_\theta & xz(1-c_\theta) + ys_\theta \\ xy(1-c_\theta) + zs_\theta & y^2(1-c_\theta) + c_\theta & yz(1-c_\theta) - xs_\theta \\ xz(1-c_\theta) - ys_\theta & yz(1-c_\theta) + xs_\theta & z^2(1-c_\theta) + c_\theta \end{bmatrix},
\end{aligned}
$$

where $c_\theta = \cos\theta$, $s_\theta = \sin\theta$, is a rotation by angle θ leaving axis u fixed.

A direction in (n+1)-dimensional space will be a unit magnitude vector, which we may consider a point on a generalized sphere, S^n. Thus it is natural to describe the rotation group SO(n+1) as combining SO(n) and S^n. A suitable formalism is the fibre bundle,

$$SO(n) \to SO(n+1) \to S^n,$$

where for every direction in the "base space", S^n, the "fibre" over it in the "total space", SO(n+1), is a copy of the "fibre space", SO(n), namely the rotations that keep that direction fixed.

Thus we can build an $n \times n$ rotation matrix by starting with a 2×2 matrix, aiming its fixed axis on S^2 (the ordinary sphere in three-dimensional space), aiming the resulting rotation on S^3, and so on up through S^{n-1}.

A point on S^n can be selected using n numbers, so we again have $n(n-1)/2$ numbers to describe any $n \times n$ rotation matrix.

In fact, we can view the sequential angle decomposition, discussed previously, as reversing this process. The composition of $n-1$ Givens rotations brings the first column (and row) to (1,0,...,0), so that the remainder of the matrix is a rotation matrix of dimension one less, embedded so as to leave (1,0,...,0) fixed.

Chapter 13

Hilbert Matrix

In linear algebra, a Hilbert matrix, introduced by Hilbert (1894), is a square matrix with entries being the unit fractions

$$H_{ij} = \frac{1}{i+j-1}.$$

For example, this is the 5 × 5 Hilbert matrix:

$$H = \begin{bmatrix} 1 & \frac{1}{2} & \frac{1}{3} & \frac{1}{4} & \frac{1}{5} \\ \text{[4pt]}\frac{1}{2} & \frac{1}{3} & \frac{1}{4} & \frac{1}{5} & \frac{1}{6} \\ \text{[4pt]}\frac{1}{3} & \frac{1}{4} & \frac{1}{5} & \frac{1}{6} & \frac{1}{7} \\ \text{[4pt]}\frac{1}{4} & \frac{1}{5} & \frac{1}{6} & \frac{1}{7} & \frac{1}{8} \\ \text{[4pt]}\frac{1}{5} & \frac{1}{6} & \frac{1}{7} & \frac{1}{8} & \frac{1}{9} \end{bmatrix}.$$

The Hilbert matrix can be regarded as derived from the integral

$$H_{ij} = \int_0^1 x^{i+j-2}\,dx,$$

that is, as a Gramian matrix for powers of x. It arises in the least squares approximation of arbitrary functions by polynomials.

The Hilbert matrices are canonical examples of ill-conditioned matrices, making them notoriously difficult to use in numerical computation. For example, the 2-norm condition number of the matrix above is about $4.8 \cdot 10^5$.

Historical Note

Hilbert (1894) introduced the Hilbert matrix to study the following question in approximation theory:

"Assume that $I = [a, b]$ is a real interval. Is it then possible to find a non-zero polynomial P with integral coefficients, such that the integral

$$\int_a^b P(x)^2\,dx$$

is smaller than any given bound $\varepsilon > 0$, taken arbitrarily small?"

To answer this question, Hilbert derives an exact formula for the determinant of the Hilbert matrices and investigates their asymptotics.

He concludes that the answer to his question is positive if the length $b - a$ of the interval is smaller than 4.

Properties

The Hilbert matrix is symmetric and positive definite.

The Hilbert matrix is also totally positive (meaning the determinant of every submatrix is positive).

The Hilbert matrix is an example of a Hankel matrix.

The determinant can be expressed in closed form, as a special case of the Cauchy determinant.

The determinant of the $n \times n$ Hilbert matrix is

$$\det(H) = \frac{c_n^{\;4}}{c_{2n}}$$

where

$$c_n = \prod_{i=1}^{n-1} i^{n-i} = \prod_{i=1}^{n-1} i!.$$

Hilbert already mentioned the curious fact that the determinant of the Hilbert matrix is the reciprocal of an integer which also follows from the identity

$$\frac{1}{\det(H)} = \frac{c_{2n}}{c_n^4} = n! \cdot \prod_{i=1}^{2n-1} \binom{i}{[i/2]}.$$

Using Stirling's approximation of the factorial one can establish the following asymptotic result:

$$\det(H) = a_n n^{-1/4} (2\pi)^n 4^{-n^2}$$

where a_n converges to the constant $e^{1/4} 2^{1/12} A^{-3} \approx 0.6450$ as $n \to \infty$, where A is the Glaisher-Kinkelin constant.

The inverse of the Hilbert matrix can be expressed in closed form using binomial coefficients; its entries are

$$(H^{-1})_{ij} = (-1)^{i+j} (i+j-1) \binom{n+i-1}{n-j} \binom{n+j-1}{n-i} \binom{i+j-2}{i-1}^2$$

where n is the order of the matrix. It follows that the entries of the inverse matrix are all integer. The condition number of the n-by-n Hilbert matrix grows as

$$O((1+\sqrt{2})^{4n} / \sqrt{n})$$

Hollow Matrix

In mathematics, a hollow matrix may refer to one of several related classes of matrix.

Sparse

A *hollow matrix* may be one with "few" non-zero entries: that is, a sparse matrix.

Diagonal Entries All Zero

A *hollow matrix* may be a square matrix whose diagonal elements are all equal to zero.

The most obvious example is the real skew-symmetric matrix.

Other examples are the adjacency matrix of a finite simple graph; a distance matrix or Euclidean distance matrix.

If A is an $n \times n$ hollow matrix, then the elements of A are given by

$$\begin{aligned} A_{n\times n} &= (a_{ij}); \\ a_{ij} &= 0 \quad \text{if} \quad i = j, \quad 1 \le i, j \le n. \end{aligned}$$

In other words, any square matrix which takes the form

$$\begin{pmatrix} 0 & & & & \\ & 0 & & & \\ & & \ddots & & \\ & & & 0 & \\ & & & & 0 \end{pmatrix}$$

is a hollow matrix.

For example:

$$\begin{pmatrix} 0 & 2 & 6 & \frac{1}{3} & 4 \\ 2 & 0 & 4 & 8 & 0 \\ 9 & 4 & 0 & 2 & 933 \\ 1 & 4 & 4 & 0 & 6 \\ 7 & 9 & 23 & 8 & 0 \end{pmatrix}$$

is an example of a hollow matrix.

Properties

- The trace of A is trivially zero.
- The linear map represented by A (with respect to a fixed basis) maps each basis vector e onto the image of the complement of $\langle e \rangle$.

Block of Zeroes

A *hollow matrix* may be a square $n \times n$ matrix with an $r \times s$ block of zeroes where $r+s>n$.

Incidence Matrix

In mathematics, an incidence matrix is a matrix that shows the relationship between two classes of objects. If the first class is X and the second is Y, the matrix has one row for each element of X and one column for each element of Y. The entry in row x and column y is 1 if x and y are related (called incident in this context) and 0 if they are not.

Graph Theory

Incidence matrices are mostly used in graph theory.

Undirected and directed graphs

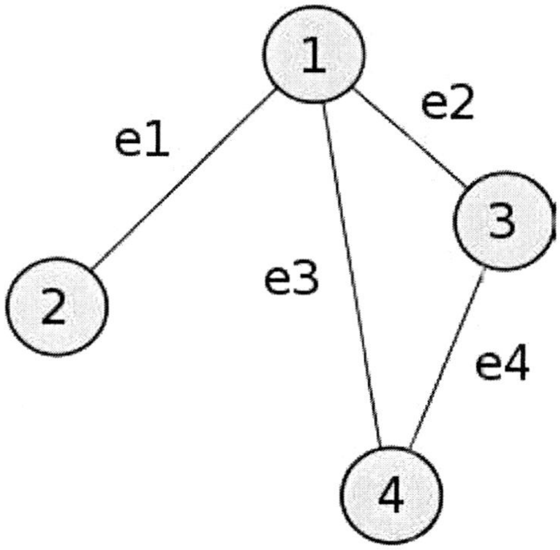

Figure: *An undirected graph.*

In graph theory an undirected graph G has two kinds of incidence matrices: unoriented and oriented. The incidence matrix (or unoriented incidence matrix) of G is a $p \times q$ matrix (b_{ij}), where p and q are the numbers of vertices and edges respectively, such that $b_{ij} = 1$ if the vertex v_i and edge x_j are incident and 0 otherwise. For example the incidence matrix of the undirected graph shown on the right is a matrix consisting of 4 rows (corresponding to the four vertices, 1-4) and 4 columns (corresponding to the four edges, e1-e4):

$$\begin{pmatrix} 1 & 1 & 1 & 0 \\ 1 & 0 & 0 & 0 \\ 0 & 1 & 0 & 1 \\ 0 & 0 & 1 & 1 \end{pmatrix}$$

If we look at the incidence matrix, we see that the sum of each column is equal to 2. This is because each edge has a vertex connected to each end.

The incidence matrix of a directed graph D is a $p \times q$ matrix $[b_{ij}]$ where p and q are the number of vertices and edges respectively, such that $b_{ij} = -1$ if the edge leaves vertex , if it enters vertex and 0 otherwise (Note that many authors use the opposite sign convention.).

An oriented incidence matrix of an undirected graph G is the incidence matrix, in the sense of directed graphs, of any orientation of G. That is, in the column of edge e, there is one +1 in the row corresponding to one vertex of e and one −1 in the row corresponding to the other vertex of e, and all other rows have 0. All oriented incidence matrices of G differ only by negating some set of columns.

In many uses, this is an insignificant difference, so one can speak of *the* oriented incidence matrix, even though that is technically incorrect.

The oriented or unoriented incidence matrix of a graph G is related to the adjacency matrix of its line graph $L(G)$ by the following theorem:

$$A(L(G)) = B(G)^T B(G) - 2I_q$$

where $A(L(G))$ is the adjacency matrix of the line graph of G, $B(G)$ is the incidence matrix, and Iq is the identity matrix of dimension q.

The Kirchhoff matrix is obtained from the oriented incidence matrix $M(G)$ by the formula

$$M(G)M(G)^T.$$

The integral cycle space of a graph is equal to the null space of its oriented incidence matrix, viewed as a matrix over the integers or real or complex numbers. The binary cycle space is the null space of its oriented or unoriented incidence matrix, viewed as a matrix over the two-element field.

Signed and Bidirected Graphs

The incidence matrix of a signed graph is a generalization of the oriented incidence matrix. It is the incidence matrix of any bidirected graph that orients the given signed graph. The column of a positive edge has a +1 in the row corresponding to one endpoint and a −1 in the row corresponding to the other endpoint, just like an edge in an ordinary (unsigned) graph. The column of a negative edge has either a +1 or a −1 in both rows. The line graph and Kirchhoff matrix properties generalize to signed graphs.

Multigraphs

The definitions of incidence matrix apply to graphs with loops and multiple edges.

The column of an oriented incidence matrix that corresponds to a loop is all zero, unless the graph is signed and the loop is negative; then the column is all zero except for ±2 in the row of its incident vertex.

Hypergraphs

Because the edges of ordinary graphs can only have two vertices (one at each end), the column of an incidence matrix for graphs can only have two non-zero entries.

By contrast, a hypergraph can have multiple vertices assigned to one edge; thus, the general case describes a hypergraph.

Incidence Structures

The incidence matrix of an incidence structure C is a $p \times q$ matrix $[b_{ij}]$, where p and q are the number of points and lines respectively, such that $b_{ij} = 1$ if the point p_i and line L_j are incident and 0 otherwise.

In this case the incidence matrix is also a biadjacency matrix of the Levi graph of the structure. As there is a hypergraph for every Levi graph, and *vice-versa*, the incidence matrix of an incidence structure describes a hypergraph.

Finite Geometries

An important example is a finite geometry. For instance, in a finite plane, X is the set of points and Y is the set of lines. In a finite geometry of higher dimension, X could be the set of points and Y could be the set of subspaces of dimension one less than the dimension of Y; or X could be the set of all subspaces of one dimension d and Y the set of all subspaces of another dimension e.

Block Designs

Another example is a block design. Here X is a finite set of "points" and Y is a class of subsets of X, called "blocks", subject to rules that depend on the type of design.

The incidence matrix is an important tool in the theory of block designs. For instance, it is used to prove the fundamental theorem of symmetric 2-designs, that the number of blocks equals the number of points.

Integer Matrix

In mathematics, an integer matrix is a matrix whose entries are all integers. Examples include binary matrices, the zero matrix, the unit matrix, and the adjacency matrices used in graph theory, amongst many others. Integer matrices find frequent application in combinatorics.

Examples

$$\begin{pmatrix} 5 & 2 & 6 & 0 \\ 4 & 7 & 3 & 8 \\ 5 & 9 & 0 & 4 \\ 3 & 1 & 0 & 3 \\ 9 & 0 & 2 & 1 \end{pmatrix} \quad \text{and} \quad \begin{pmatrix} 1 & 5 & 0 \\ 0 & 9 & 2 \\ 1 & 7 & 3 \end{pmatrix}$$

are both examples of integer matrices

Invertible Matrix

In linear algebra an n-by-n (square) matrix A is called invertible (some authors use nonsingular or nondegenerate) if there exists an n-by-n matrix B such that

$$\mathbf{AB} = \mathbf{BA} = \mathbf{I}_n$$

where I_n denotes the n-by-n identity matrix and the multiplication used is ordinary matrix multiplication. If this is the case, then the matrix B is uniquely determined by A and is called the *inverse* of A, denoted by A^{-1}. It follows from the theory of matrices that if

$$\mathbf{AB} = \mathbf{I}$$

for *finite square* matrices A and B, then also

$$\mathbf{BA} = \mathbf{I}.$$

Non-square matrices (m-by-n matrices for which $m \neq n$) do not have an inverse. However, in some cases such a matrix may have a left inverse or right inverse. If A is m-by-n and the rank of A is equal to n, then A has a left inverse: an n-by-m

matrix B such that BA = I. If A has rank m, then it has a right inverse: an n-by-m matrix B such that AB = I.

A square matrix that is not invertible is called singular or degenerate. A square matrix is singular if and only if its determinant is 0. Singular matrices are rare in the sense that if you pick a random square matrix over a continuous uniform distribution on its entries, it will almost surely not be singular.

While the most common case is that of matrices over the real or complex numbers, all these definitions can be given for matrices over any commutative ring.

However, in this case the condition for a square matrix to be invertible is that its determinant is invertible in the ring, which in general is a much stricter requirement than being nonzero.

The conditions for existence of left-inverse resp. right-inverse are more complicated since a notion of rank does not exist over rings.

Matrix inversion is the process of finding the matrix B that satisfies the prior equation for a given invertible matrix A.

Properties

Let A be a square n by n matrix over a field K (for example the field R of real numbers). The following statements are equivalent:

A is invertible.

A is row-equivalent to the n-by-n identity matrix I_n.

A is column-equivalent to the n-by-n identity matrix I_n.

A has n pivot positions.

det A $\neq$ 0. In general, a square matrix over a commutative ring is invertible if and only if its determinant is a unit in that ring.

rank A = n.

The equation Ax = 0 has only the trivial solution x = 0 (i.e., Null A = {0})

The equation Ax = b has exactly one solution for each b in K^n, (x $\neq$ 0).

The columns of A are linearly independent.

The columns of A span K^n (i.e. Col A = K^n).

The columns of A form a basis of K^n.

The linear transformation mapping x to Ax is a bijection from K^n to K^n.

There is an n by n matrix B such that AB = I_n = BA.

The transpose A^T is an invertible matrix (hence rows of A are linearly independent, span K^n, and form a basis of K^n).

The number 0 is not an eigenvalue of A.

The matrix A can be expressed as a finite product of elementary matrices.

Furthermore, the following properties hold for an invertible matrix A:

- $(A^{-1})^{-1} = A$;
- $(kA)^{-1} = k^{-1}A^{-1}$ for nonzero scalar k;
- $(A^T)^{-1} = (A^{-1})^T$;
- For any invertible n-by-n matrices A and B, $(AB)^{-1} = B^{-1}A^{-1}$. More generally, if $A_1,\ldots,A_k$ are invertible n-by-n matrices, then $(A_1A_2$ï–$A_{k-1}A_k)^{-1} = A_k^{-1}A_{k-1}^{-1}$ï–$A_2^{-1}A_1^{-1}$;
- $\det(A^{-1}) = \det(A)^{-1}$.

A matrix that is its own inverse, i.e. $A = A^{-1}$ and $A^2 = I$, is called an involution.

Density

Over the field of real numbers, the set of singular n-by-n matrices, considered as a subset of $R^{n\times n}$, is a null set, i.e., has Lebesgue measure zero. This is true because singular matrices are the roots of the polynomial function in the entries of the matrix given by the determinant. Thus in the language of measure theory, almost all n-by-n matrices are invertible.

Furthermore the n-by-n invertible matrices are a dense open set in the topological space of all n-by-n matrices. Equivalently, the set of singular matrices is closed and nowhere dense in the space of n-by-n matrices.

In practice however, one may encounter non-invertible matrices. And in numerical calculations, matrices which are invertible, but close to a non-invertible matrix, can still be problematic; such matrices are said to be ill-conditioned.

Methods of Matrix Inversion

Gaussian Elimination

Gauss–Jordan elimination is an algorithm that can be used to determine whether a given matrix is invertible and to find the inverse. An alternative is the LU decomposition which generates upper and lower triangular matrices which are easier to invert.

For special purposes, it may be convenient to invert matrices by treating $m \cdot n$-by-$m \cdot n$ matrices as m-by-m matrices of n-by-n matrices, and applying one or another formula recursively (other sized matrices can be padded out with dummy rows and columns).

For other purposes, a variant of Newton's method may be convenient. Newton's method is particularly useful when dealing with families of related matrices: sometimes a good starting point for refining an approximation for the new inverse can be the already obtained inverse of a previous matrix that nearly matches the current matrix. Newton's method is also useful for "touch up" corrections to the Gauss–Jordan algorithm which has been contaminated by small errors due to imperfect computer arithmetic.

Eigendecomposition

If matrix A can be eigendecomposed and if none of its eigenvalues are zero, then A is nonsingular and its inverse is given by

$$\mathbf{A}^{-1} = \mathbf{Q}\Lambda^{-1}\mathbf{Q}^{-1}$$

Furthermore, because $\wedge$ is a diagonal matrix, its inverse is easy to calculate:

$$\left[\Lambda^{-1}\right]_{ii} = \frac{1}{\lambda_i}$$

Cholesky Decomposition

If matrix A is positive definite, then its inverse can be obtained as

$$\mathbf{A}^{-1} = (\mathbf{L}^*)^{-1}\mathbf{L}^{-1},$$

where L is the lower triangular Cholesky decomposition of A.

Analytic Solution

Writing the transpose of the matrix of cofactors, known as an adjugate matrix, can also be an efficient way to calculate the inverse of *small* matrices, but this recursive method is inefficient for large matrices. To determine the inverse, we calculate a matrix of cofactors:

$$\mathbf{A}^{-1} = \frac{1}{|\mathbf{A}|}\left(\mathbf{C}^{\mathrm{T}}\right)_{ij} = \frac{1}{|\mathbf{A}|}\left(\mathbf{C}_{ji}\right) = \frac{1}{|\mathbf{A}|}\begin{pmatrix} \mathbf{C}_{11} & \mathbf{C}_{21} & \cdots & \mathbf{C}_{n1} \\ \mathbf{C}_{12} & \mathbf{C}_{22} & \cdots & \mathbf{C}_{n2} \\ \vdots & \vdots & \ddots & \vdots \\ \mathbf{C}_{1n} & \mathbf{C}_{2n} & \cdots & \mathbf{C}_{nn} \end{pmatrix}$$

where $|A|$ is the determinant of A, C_{ij} is the matrix of cofactors, and C^T represents the matrix transpose.

Inversion of 2×2 matrices

The *cofactor equation* listed above yields the following result for 2×2 matrices. Inversion of these matrices can be done easily as follows:

$$\mathbf{A}^{-1} = \begin{bmatrix} a & b \\ c & d \end{bmatrix}^{-1} = \frac{1}{\det(\mathbf{A})}\begin{bmatrix} d & -b \\ -c & a \end{bmatrix} = \frac{1}{ad-bc}\begin{bmatrix} d & -b \\ -c & a \end{bmatrix}.$$

This is possible because 1/(ad-bc) is the reciprocal of the determinant of the matrix in question, and the same strategy could be used for other matrix sizes.

Inversion of 3×3 matrices

A computationally efficient 3x3 matrix inversion is given by

$$\mathbf{A}^{-1} = \begin{bmatrix} a & b & c \\ d & e & f \\ g & h & k \end{bmatrix}^{-1} = \frac{1}{\det(\mathbf{A})}\begin{bmatrix} A & B & C \\ D & E & F \\ G & H & K \end{bmatrix}^{T}$$

$$= \frac{1}{\det(\mathbf{A})}\begin{bmatrix} A & D & G \\ B & E & H \\ C & F & K \end{bmatrix}$$

where the determinant of A can be computed by applying the rule of Sarrus as follows:

$$\det(\mathbf{A}) = a(ek - fh) - b(kd - fg) + c(dh - eg).$$

If the determinant is non-zero, the matrix is invertible, with the elements of the above matrix on the right side given by

$$A = (ek - fh) \quad D = (ch - bk) \quad G = (bf - ce)$$
$$B = (fg - dk) \quad E = (ak - cg) \quad H = (cd - af)$$
$$C = (dh - eg) \quad F = (gb - ah) \quad K = (ae - bd).$$

The general 3×3 inverse can be expressed concisely in terms of the cross product and triple product:

If a matrix $\mathbf{A} = [\mathbf{x}_0, \mathbf{x}_1, \mathbf{x}_2]$ (consisting of three column vectors, $\mathbf{x}_0, \mathbf{x}_1,$ and $\mathbf{x}_2$) is invertible, its inverse is given by

$$\mathbf{A}^{-1} = \frac{1}{\det(\mathbf{A})} \begin{bmatrix} (\mathbf{x}_1 \times \mathbf{x}_2)^T \\ (\mathbf{x}_2 \times \mathbf{x}_0)^T \\ (\mathbf{x}_0 \times \mathbf{x}_1)^T \end{bmatrix}$$

Note that $\det(A)$ is equal to the triple product of X_0, X_1, and X_2—the volume of the parallelepiped formed by the rows or columns:

$$\det(\mathbf{A}) = \mathbf{x}_0 \cdot (\mathbf{x}_1 \times \mathbf{x}_2).$$

The correctness of the formula can be checked by using cross- and triple-product properties and by noting that for groups, left and right inverses always coincide. Intuitively, because of the cross products, each row of $\mathbf{A}^{-1}$ is orthogonal to the non-corresponding two columns of A(causing the off-diagonal terms of $I = \mathbf{A}^{-1}\mathbf{A}$ be zero). Dividing by

$$\det(\mathbf{A}) = \mathbf{x}_0 \cdot (\mathbf{x}_1 \times \mathbf{x}_2)$$

causes the diagonal elements of $I = \mathbf{A}^{-1}\mathbf{A}$ to be unity. For example, the first diagonal is:

$$1 = \frac{1}{\mathbf{x}_0 \cdot (\mathbf{x}_1 \times \mathbf{x}_2)} \mathbf{x}_0 \cdot (\mathbf{x}_1 \times \mathbf{x}_2).$$

Moore–Penrose Pseudoinverse

Some of the properties of inverse matrices are shared by Moore–Penrose pseudoinverses, which can be defined for any m-by-n matrix.

Applications

For most practical applications, it is *not* necessary to invert a matrix to solve a system of linear equations; however, for a unique solution, it *is* necessary that the matrix involved be invertible. Decomposition techniques like LU decomposition are much faster than inversion, and various fast algorithms for special classes of linear systems have also been developed.

Matrix Inverses in Real-Time Simulations

Matrix inversion plays a significant role in computer graphics, particularly in 3D graphics rendering and 3D simulations. Examples include screen-to-world ray casting, world-to-subspace-to-world object transformations, and physical simulations.

Involutory Matrix

In mathematics, an involutory matrix is a matrix that is its own inverse. That is, matrix A is an involution if and only if $A^2 = I$. One of the three classes of elementary matrix is involutory, namely the *row-interchange elementary matrix*. A special case of another class of elementary matrix, that which represents multiplication of a row or column by -1, is also involutory; it is in fact a trivial example of a signature matrix, all of which are involutory.

Involutory matrices are all square roots of the identity matrix. This is simply a consequence of the fact that any nonsingular matrix multiplied by its inverse is the identity. If A is an $n \times n$ matrix, then A is involutory if and only if $\frac{1}{2}(A + I)$ is idempotent.

An involutory matrix which is also symmetric is an orthogonal matrix, and thus represents an isometry (a linear transformation which preserves Euclidean distance). A reflection matrix is an example of an involutory matrix.

Clearly, any block-diagonal matrices constructed from involutory matrices will also be involutory, as a consequence of the linear independence of the blocks.

Examples

Some simple examples of involutory matrices are shown below.

$$\mathbf{I} = \begin{pmatrix} 1 & 0 & 0 \\ 0 & 1 & 0 \\ 0 & 0 & 1 \end{pmatrix}; \qquad \mathbf{I}^{-1} = \begin{pmatrix} 1 & 0 & 0 \\ 0 & 1 & 0 \\ 0 & 0 & 1 \end{pmatrix}$$

$$\mathbf{R} = \begin{pmatrix} 1 & 0 & 0 \\ 0 & 0 & 1 \\ 0 & 1 & 0 \end{pmatrix}; \qquad \mathbf{R}^{-1} = \begin{pmatrix} 1 & 0 & 0 \\ 0 & 0 & 1 \\ 0 & 1 & 0 \end{pmatrix}$$

$$\mathbf{S} = \begin{pmatrix} +1 & 0 & 0 \\ 0 & -1 & 0 \\ 0 & 0 & -1 \end{pmatrix}; \quad \mathbf{S}^{-1} = \begin{pmatrix} +1 & 0 & 0 \\ 0 & -1 & 0 \\ 0 & 0 & -1 \end{pmatrix}$$

where

I is the identity matrix (which is trivially involutory);

R is a matrix with a pair of interchanged rows;

S is a signature matrix.

An interesting general condition exists, for 2 × 2 matrices, where any matrix that may be written in the form A or A^T below:

$$\mathbf{A} = \begin{pmatrix} a & b \\ \dfrac{(1-a^2)}{b} & -a \end{pmatrix}; \quad \mathbf{A}^T = \begin{pmatrix} a & \dfrac{(1-a^2)}{b} \\ b & -a \end{pmatrix}$$

is involutory.

For example, for a matrix M of this form, where we set $a = 1$, $b = 1$, we have

$$\mathbf{M} = \begin{pmatrix} 1 & 1 \\ 0 & -1 \end{pmatrix}; \quad \Rightarrow \quad \mathbf{M}^2 = \begin{pmatrix} 1\times1+1\times0 & 1\times1+1\times-1 \\ 0\times1-1\times0 & 0\times1-1\times-1 \end{pmatrix}$$

$$= \begin{pmatrix} 1 & 0 \\ 0 & 1 \end{pmatrix} = \mathbf{I}$$

Irregular Matrix

An irregular matrix, or ragged matrix, can be described as a matrix that has a different number of elements in each row. Ragged matrices are not used in linear algebra, since standard matrix transformations cannot be performed on them, but they are useful as arrays in computing. Irregular matrices are typically stored using Iliffe vectors.

For example, the following is an irregular matrix:

$$\begin{bmatrix} 1 & 31 & 12 & -3 \\ 7 & 2 & & \\ 1 & 2 & 2 & \end{bmatrix}$$

Leslie Matrix

In applied mathematics, the Leslie matrix is a discrete, age-structured model of population growth that is very popular in population ecology. It was invented by and named after Patrick H. Leslie. The Leslie matrix (also called the Leslie Model) is one of the best known ways to describe the growth of populations (and their projected age distribution), in which a population is closed to migration and where only one sex, usually the female, is considered.

The Leslie Matrix is used in ecology to model the changes in a population of organisms over a period of time.

In a Leslie Model, the population is divided into groups based on age classes. A similar model which replaces age classes with life stage is called a Lefkovitch matrix, whereby individuals can both remain in the same stage class or move on to the next one. At each time step the population is represented by a vector with an element for each age classes where each element indicates the number of individuals currently in that class.

The Leslie Matrix is a square matrix with the same number of rows and columns as the population vector has elements.

The (i,j)th cell in the matrix indicates how many individuals will be in the age class i at the next time step for each individual in stage j. At each time step, the population vector is multiplied by the Leslie Matrix to generate the population vector for the following time step.

To build a matrix, some information must be known from the population:

- n_x, the number of individual (n) of each age class x

- s_x, the fraction of individuals that survives from age class x to age class $x+1$,
- f_x, fecundity, the per capita average number of female offspring reaching n_0 born from mother of the age class x More precisely it can be viewed as the number of offspring produced at the next age class b_{x+1} weighted by the probability of reaching the next age class. Therefore $f_x = s_x b_{x+1}$

The observations that at time *t+1* is simply the sum of all offspring born from the previous time step and that the organisms surviving to time *t+1* are the organisms at time *t* surviving at probability we get This then motivates the following matrix representation:

$$\begin{bmatrix} n_0 \\ n_1 \\ \vdots \\ n_{\omega-1} \end{bmatrix}_{t+1} = \begin{bmatrix} f_0 & f_1 & f_2 & f_3 & \cdots & f_{\omega-1} \\ s_0 & 0 & 0 & 0 & \cdots & 0 \\ 0 & s_1 & 0 & 0 & \cdots & 0 \\ 0 & 0 & s_2 & 0 & \cdots & 0 \\ 0 & 0 & 0 & \ddots & \cdots & 0 \\ 0 & 0 & 0 & \cdots & s_{\omega-2} & 0 \end{bmatrix} \begin{bmatrix} n_0 \\ n_1 \\ \vdots \\ n_{\omega-1} \end{bmatrix}_t$$

Where ω is the maximum age attainable in our population. This can be written as;

$$\mathbf{n}_{t+1} = \mathbf{L}\mathbf{n}_t$$

or;

$$\mathbf{n}_t = \mathbf{L}^t\mathbf{n}_0$$

Where n_t is the population vector at time t and L is the Leslie matrix. The dominant eigenvalue of , denoted , gives the populations asymptotic growth rate (growth rate at the stable age distribution). The corresponding eigenvector provides the stable age distribution, the proportion of individuals of each age within the population. Once the stable age distribution has been reached, a population undergoes exponential growth at rate .

The characteristic polynomial of the matrix is given by the Euler–Lotka equation. The Leslie model is very similar to a discrete-time Markov chain. The main difference is that in a Markov model, one would have $f_x + s_x = 1$ for each , while the Leslie model may have these sums greater or less than 1.

Stable Age Structure

This age-structured growth model suggests a steady-state, or stable, age-structure and growth rate. Regardless of the initial population size, N_0, or age distribution, the population tends asymptotically to this age-structure and growth rate. It also returns to this state following perturbation. The Euler–Lotka equation provides a means of identifying the intrinsic growth rate.

The stable age-structure is determined both by the growth rate and the survival function (i.e. the Leslie matrix). For example, a population with a large intrinsic growth rate will have a disproportionately "young" age-structure. A population with high mortality rates at all ages (i.e. low survival) will have a similar age-structure. Charlesworth (1980) provides further details on the rate and form of convergence to the stable age-structure.

Levinson Recursion

Levinson recursion or Levinson–Durbin recursion is a procedure in linear algebra to recursively calculate the solution to an equation involving a Toeplitz matrix. The algorithm runs in $\Theta(n^2)$ time, which is a strong improvement over Gauss–Jordan elimination, which runs in $\Theta(n^3)$.

Newer algorithms, called *asymptotically fast* or sometimes *superfast* Toeplitz algorithms, can solve in $\Theta(n \log^p n)$ for various p (e.g. p = 2, p = 3). Levinson recursion remains popular for several reasons; for one, it is relatively easy to understand in comparison; for another, it can be faster than a superfast algorithm for small n (usually n < 256).

The Levinson-Durbin algorithm was proposed first by Norman Levinson in 1947, improved by James Durbin in 1960, and subsequently improved to $4n^2$ and then $3n^2$ multiplications by W. F. Trench and S. Zohar, respectively.

Other methods to process data include Schur decomposition and Cholesky decomposition. In comparison to these, Levinson recursion (particularly Split-Levinson recursion) tends to be faster computationally, but more sensitive to computational inaccuracies like round-off errors.

Derivation

Background

Matrix equations follow the form:

$$\vec{y} = \mathbf{M}\,\vec{x}.$$

The Levinson-Durbin algorithm may be used for any such equation, as long as M is a known Toeplitz matrix with a nonzero main diagonal. Here $\vec{y}$ is a known vector, and $\vec{x}$ is an unknown vector of numbers x_i yet to be determined.

For the sake of this article, κ_i is a vector made up entirely of zeroes, except for its i'th place, which holds the value one. Its length will be implicitly determined by the surrounding context. The term N refers to the width of the matrix above — M is an $N \times N$ matrix.

Finally, in this article, superscripts refer to an *inductive index*, whereas subscripts denote indices. For example (and definition), in this article, the matrix T^n is an $n \times n$ matrix which copies the upper left $n \times n$ block from M — that is, $T^n_{ij} = M_{ij}$.

T^n is also a Toeplitz matrix; meaning that it can be written as:

$$\mathbf{T}^n = \begin{bmatrix} t_0 & t_{-1} & t_{-2} & \cdots & t_{-n+1} \\ t_1 & t_0 & t_{-1} & \cdots & t_{-n+2} \\ t_2 & t_1 & t_0 & \cdots & t_{-n+3} \\ \vdots & \vdots & \vdots & \ddots & \vdots \\ t_{n-1} & t_{n-2} & t_{n-3} & \cdots & t_0 \end{bmatrix}.$$

Introductory Steps

The algorithm proceeds in two steps. In the first step, two sets of vectors, called the *forward* and *backward* vectors, are established.

The forward vectors are used to help get the set of backward vectors; then they can be immediately discarded. The backwards vectors are necessary for the second step, where they are used to build the solution desired.

Levinson-Durbin recursion defines the nth "forward vector", denoted $\vec{f}^n$, as the vector of length n which satisfies:

$$\mathbf{T}^n \vec{f}^n = \hat{e}_1.$$

The nth "backward vector" $\vec{b}^n$ is defined similarly; it is the vector of length n which satisfies:

$$\mathbf{T}^n \vec{b}^n = \hat{e}_n.$$

An important simplification can occur when M is a symmetric matrix; then the two vectors are related by $b^n_i = f^n_{n+1-i}$ — that is, they are row-reversals of each other. This can save some extra computation in that special case.

Obtaining the Backward Vectors

Even if the matrix is not symmetric, then the nth forward and backward vector may be found from the vectors of length n-1 as follows. First, the forward vector may be extended with a zero to obtain:

$$\mathbf{T}^n \begin{bmatrix} \vec{f}^{n-1} \\ 0 \end{bmatrix} = \begin{bmatrix} & & & t_{-n+1} \\ & \mathbf{T}^{n-1} & & t_{-n+2} \\ & & & \vdots \\ t_{n-1} & t_{n-2} & \dots & t_0 \end{bmatrix} \begin{bmatrix} \vec{f}^{n-1} \\ \\ 0 \end{bmatrix} = \begin{bmatrix} 1 \\ 0 \\ \vdots \\ 0 \\ \epsilon_f^n \end{bmatrix}.$$

In going from T^{n-1} to T^n, the extra *column* added to the matrix does not perturb the solution when a zero is used to extend the forward vector.

However, the extra *row* added to the matrix *has* perturbed the solution; and it has created an unwanted error term ε_f which occurs in the last place. The above equation gives it the value of:

$$\epsilon_f^n = \sum_{i=1}^{n-1} M_{ni}\, f_i^{n-1} = \sum_{i=1}^{n-1} t_{n-i}\, f_i^{n-1}.$$

This error will be returned to shortly and eliminated from the new forward vector; but first, the backwards vector must be extended in a similar (albeit reversed) fashion. For the backwards vector,

$$\mathbf{T}^n \begin{bmatrix} 0 \\ \vec{b}^{n-1} \end{bmatrix} = \begin{bmatrix} t_0 & \cdots & t_{-n+2} & t_{-n+1} \\ \vdots & & & \\ t_{n-2} & & \mathbf{T}^{n-1} & \\ t_{n-1} & & & \end{bmatrix} \begin{bmatrix} 0 \\ \\ \vec{b}^{n-1} \\ \\ \end{bmatrix} = \begin{bmatrix} \epsilon_b^n \\ 0 \\ \vdots \\ 0 \\ 1 \end{bmatrix}.$$

Like before, the extra column added to the matrix does not perturb this new backwards vector; but the extra row does. Here we have another unwanted error ε_b with value:

$$\epsilon_b^n = \sum_{i=2}^{n} M_{1i}\, b_{i-1}^{n-1} = \sum_{i=1}^{n-1} t_{-i}\, b_i^{n-1}.$$

These two error terms can be used to eliminate each other. Using the linearity of matrices,

$$\forall(\alpha,\beta)\ \mathbf{T}\left(\alpha \begin{bmatrix} \vec{f} \\ \\ 0 \end{bmatrix} + \beta \begin{bmatrix} 0 \\ \\ \vec{b} \end{bmatrix} \right) = \alpha \begin{bmatrix} 1 \\ 0 \\ \vdots \\ 0 \\ \epsilon_f \end{bmatrix} + \beta \begin{bmatrix} \epsilon_b \\ 0 \\ \vdots \\ 0 \\ 1 \end{bmatrix}.$$

If α and β are chosen so that the right hand side yields κ_1 or κ_n, then the quantity in the parentheses will fulfill the definition of the nth forward or backward vector, respectively.

With those alpha and beta chosen, the vector sum in the parentheses is simple and yields the desired result.

To find these coefficients, α_f^n, β_f^n are such that :

$$\vec{f}_n = \alpha_f^n \begin{bmatrix} \vec{f}_{n-1} \\ 0 \end{bmatrix} + \beta_f^n \begin{bmatrix} 0 \\ \vec{b}_{n-1} \end{bmatrix}$$

and respectively α_b^n, β_b^n are such that :

$$\vec{b}_n = \alpha_b^n \begin{bmatrix} \vec{f}_{n-1} \\ 0 \end{bmatrix} + \beta_b^n \begin{bmatrix} 0 \\ \vec{b}_{n-1} \end{bmatrix}.$$

By multiplying both previous equations by T^n one gets the following equation:

$$\begin{bmatrix} 1 & \epsilon_b^n \\ 0 & 0 \\ \vdots & \vdots \\ 0 & 0 \\ \epsilon_f^n & 1 \end{bmatrix} \begin{bmatrix} \alpha_f^n & \alpha_b^n \\ \beta_f^n & \beta_b^n \end{bmatrix} = \begin{bmatrix} 1 & 0 \\ 0 & 0 \\ \vdots & \vdots \\ 0 & 0 \\ 0 & 1 \end{bmatrix}.$$

Now, all the zeroes in the middle of the two vectors above being disregarded and collapsed, only the following equation is left:

$$\begin{bmatrix} 1 & \epsilon_b^n \\ \epsilon_f^n & 1 \end{bmatrix} \begin{bmatrix} \alpha_f^n & \alpha_b^n \\ \beta_f^n & \beta_b^n \end{bmatrix} = \begin{bmatrix} 1 & 0 \\ 0 & 1 \end{bmatrix}.$$

With these solved for (by using the Cramer 2x2 matrix inverse formula), the new forward and backward vectors are:

$$\vec{f}^n = \frac{1}{1-\epsilon_b^n \epsilon_f^n} \begin{bmatrix} \vec{f}^{n-1} \\ 0 \end{bmatrix} - \frac{\epsilon_f^n}{1-\epsilon_b^n \epsilon_f^n} \begin{bmatrix} 0 \\ \vec{b}^{n-1} \end{bmatrix}$$

$$\vec{b}^n = \frac{1}{1-\epsilon_b^n \epsilon_f^n} \begin{bmatrix} 0 \\ \vec{b}^{n-1} \end{bmatrix} - \frac{\epsilon_b^n}{1-\epsilon_b^n \epsilon_f^n} \begin{bmatrix} \vec{f}^{n-1} \\ 0 \end{bmatrix}.$$

Performing these vector summations, then, gives the nth forward and backward vectors from the prior ones. All that remains is to find the first of these vectors, and then some quick sums and multiplications give the remaining ones. The first forward and backward vectors are simply:

$$\vec{f}^1 = \vec{b}^1 = \left[\frac{1}{M_{11}}\right] = \left[\frac{1}{t_0}\right].$$

Using the backward vectors

Using the Backward Vectors

The above steps give the N backward vectors for M. From there, a more arbitrary equation is:

$$\vec{y} = \mathbf{M}\,\vec{x}.$$

The solution can be built in the same recursive way that the backwards vectors were built. Accordingly, $\vec{x}$ must be generalized to a sequence $\vec{x}^n$, from which $\vec{x}^N = \vec{x}$.

The solution is then built recursively by noticing that if:

$$\mathbf{T}^{n-1}\begin{bmatrix} x_1^{n-1} \\ x_2^{n-1} \\ \dots \\ x_{n-1}^{n-1} \end{bmatrix} = \begin{bmatrix} y_1 \\ y_2 \\ \dots \\ y_{n-1} \end{bmatrix}.$$

Then, extending with a zero again, and defining an error constant where necessary:

$$\mathbf{T}^{n}\begin{bmatrix} x_1^{n-1} \\ x_2^{n-1} \\ \dots \\ x_{n-1}^{n-1} \\ 0 \end{bmatrix} = \begin{bmatrix} y_1 \\ y_2 \\ \dots \\ y_{n-1} \\ \epsilon_x^{n-1} \end{bmatrix}.$$

We can then use the nth backward vector to eliminate the error term and replace it with the desired formula as follows:

$$\mathbf{T}^{n}\left(\begin{bmatrix} x_1^{n-1} \\ x_2^{n-1} \\ \dots \\ x_{n-1}^{n-1} \\ 0 \end{bmatrix} + (y_n - \epsilon_x^{n-1})\,\vec{b}^n\right) = \begin{bmatrix} y_1 \\ y_2 \\ \dots \\ y_{n-1} \\ y_n \end{bmatrix}.$$

Extending this method until n = N yields the solution $\vec{x}$.

In practice, these steps are often done concurrently with the rest of the procedure, but they form a coherent unit and deserve to be treated as their own step.

Block Levinson Algorithm

If *M* is not strictly Toeplitz, but block Toeplitz, the Levinson recursion can be derived in much the same way by regarding the block Toeplitz matrix as a Toeplitz matrix with matrix elements (Musicus 1988).

Block Toeplitz matrices arise naturally in signal processing algorithms when dealing with multiple signal streams (e.g., in MIMO systems) or cyclo-stationary signals.

Logical Matrix

A logical matrix, binary matrix, relation matrix, Boolean matrix, or (0,1) matrix is a matrix with entries from the Boolean domain B = {0, 1}. Such a matrix can be used to represent a binary relation between a pair of finite sets.

Matrix Representation of a Relation

If R is a binary relation between the finite indexed sets X and Y (so R †– $X \times Y$), then R can be represented by the adjacency matrix M whose row and column indices index the elements of X and Y, respectively, such that the entries of M are defined by:

$$M_{i,j} = \begin{cases} 1 & (x_i, y_j) \in R \\ 0 & (x_i, y_j) \notin R \end{cases}$$

In order to designate the row and column numbers of the matrix, the sets X and Y are indexed with positive integers: i ranges from 1 to the cardinality of X and j ranges from 1 to the cardinality of Y.

Example

The binary relation R on the set {1, 2, 3, 4} is defined so that aRb holds if and only if a divides b evenly, with no remainder. For example, 2R4 holds because 2 divides 4 without leaving a remainder, but 3R4 does not hold because when 3 divides 4 there is a remainder of 1. The following set is the set of pairs for which the relation R holds.

{(1, 1), (1, 2), (1, 3), (1, 4), (2, 2), (2, 4), (3, 3), (4, 4)}.

The corresponding representation as a Boolean matrix is:

$$\begin{pmatrix} 1 & 1 & 1 & 1 \\ 0 & 1 & 0 & 1 \\ 0 & 0 & 1 & 0 \\ 0 & 0 & 0 & 1 \end{pmatrix}.$$

Other Examples

- A permutation matrix is a (0,1)-matrix, all of whose columns and rows each have exactly one nonzero element.

 - o A Costas array is a special case of a permutation matrix
- An incidence matrix in combinatorics and finite geometry has ones to indicate incidence between points (or vertices) and lines of a geometry, blocks of a block design, or edges of a graph (mathematics)
- A design matrix in analysis of variance is a (0,1)-matrix with constant row sums.
- An adjacency matrix in graph theory is a matrix whose rows and columns represent the vertices and whose entries represent the edges of the graph. The adjacency matrix of a simple, undirected graph is a binary symmetric matrix with zero diagonal.
- The biadjacency matrix of a simple, undirected bipartite graph is a (0,1)-matrix, and any (0,1)-matrix arises in this way.
- The prime factors of a list of m square-free, n-smooth numbers can be described as a $m \times \pi(n)$ (0,1)-matrix, where π is the prime-counting function and a_{ij} is 1 if and only if the jth prime divides the ith number. This representation is useful in the quadratic sieve factoring algorithm.
- A bitmap image containing pixels in only two colours can be represented as a (0,1)-matrix in which the 0's represent pixels of one colour and the 1's represent pixels of the other colour.
- Using binary matrix to check the game rules in the game of Go [1]

Some Properties

The matrix representation of the equality relation on a finite set is an identity matrix, that is, one whose entries on the diagonal are all 1, while the others are all 0. If the Boolean domain is viewed as a semiring, where addition corresponds to logical OR and multiplication to logical AND, the matrix representation of the composition of two relations is equal to the matrix product of the matrix representations of these relation. Frequently operations on binary matrices are defined in terms of modular arithmetic mod 2 — that is, the elements are treated as elements of the Galois field GF(2) = $\$!_2$. They arise in a variety of representations and have a number of more restricted special forms. The number of distinct m-by-n binary matrices is equal to 2^{mn}, and is thus finite.

M-matrix

In mathematics, especially linear algebra, an *M*-matrix is a *Z*-matrix with eigenvalues whose real parts are positive. *M*-matrices are a subset of the class of *P*-matrices, and also of the class of inverse-positive matrices (i.e. matrices with inverses belonging to the class of positive matrices).

A common characterization of an *M*-matrix is a non-singular square matrix with non-positive off-diagonal entries and all principal minors positive, but many equivalences are known. The name *M*-matrix was seemingly originally chosen by Alexander Ostrowski in reference to Hermann Minkowski.

A symmetric *M*-matrix is sometimes called a Stieltjes matrix.

M-matrices arise naturally in some discretizations of differential operators, particularly those with a minimum/ maximum principle, such as the Laplacian, and as such are well-studied in scientific computing.

The *LU* factors of an *M*-matrix are guaranteed to exist and can be stably computed without need for numerical pivoting, also have positive diagonal entries and non-positive off-diagonal entries. Furthermore, this holds even for incomplete *LU* factorization, where entries in the factors are discarded during factorization, providing useful preconditioners for iterative solution.

Bibliography

Arnold, Vladimir I.; Cooke, Roger: *Ordinary Differential Equations*, Springer-Verlag, Berlin, DE; New York, 2012.

Bau III, David; Trefethen, Lloyd N.: *Numerical Linear Algebra*, Society for Industrial and Applied Mathematics, Philadelphia, PA, 2017.

Bretscher, Otto: *Linear Algebra with Applications,* Prentice Hall, NJ, 2015.

Bronson, Richard: *Schaum's Outline of Theory and Problems of Matrix Operations*, McGraw–Hill, New York, 2009.

Brown, William A.: *Matrices and Vector Spaces*, M. Dekker, New York, 2011.

Coburn, Nathaniel: *Vector and Tensor Analysis*, Macmillan, New York, 1995.

Conrey, J. Brian: *Ranks of Elliptic Curves and Random Matrix Theory*, Cambridge University Press, 2017.

Gilbarg, David; Trudinger, Neil S.: *Elliptic Partial Differential Equations of Second Order,* Springer-Verlag, Berlin, DE; New York, 2023.

Godsil, Chris; Royle, Gordon: *Algebraic Graph Theory*, Graduate Texts in Mathematics, Springer-Verlag, Berlin, DE; New York, 2014.

Golub, Gene H.; Van Loan, Charles F.: *Matrix Computations,* Johns Hopkins, 2016.

Greub, Werner Hildbert: *Linear Algebra*, Graduate Texts in Mathematics, Springer-Verlag, Berlin, DE; New York, 2005.

Halmos, Paul Richard: *A Hilbert Space Problem Book*, Graduate Texts in Mathematics, Springer-Verlag, Berlin, DE; New York, 2002.

Horn, Roger A.; Johnson, Charles R.: *Matrix Analysis*, Cambridge University Press, 2005.

Householder, Alston S.: *The Theory of Matrices in Numerical Analysis*, Dover Publications, New York, 2005.

Krzanowski, Wojtek J.: *Principles of Multivariate Analysis*, The Clarendon Press Oxford University Press, 2008.

Lang, Serge: *Calculus of Several Variables,* Berlin, Springer-Verlag, DE; New York, 2007.

____________: *Analysis II*, Addison-Wesley, USA, 2009.

Latouche, Guy; Ramaswami, Vaidyanathan: *Introduction to Matrix Analytic Methods in Stochastic Modeling,* Society for Industrial and Applied Mathematics, Philadelphia, PA, 2019.

Manning, Christopher D.; Schütze, Hinrich: *Foundations of Statistical Natural Language Processing*, MIT Press, 2018.

Mehata, K. M.; Srinivasan, S. K.: *Stochastic Processes*, McGraw–Hill, New York, 2008.

Nocedal, Jorge; Wright, Stephen J.: *Numerical Optimization,* Springer-Verlag, Berlin, DE; New York, 2016.

Oualline, Steve: *Practical C++ Programming*, O'Reilly, 2013.

Press, William H.; Flannery, Brian P.; Teukolsky, Saul A.; Vetterling, William T.: "LU Decomposition and Its Applications", *Numerical Recipes in FORTRAN: The Art of Scientific Computing*, Cambridge University Press, 2012.

Punnen, Abraham P.; Gutin, Gregory: *The Traveling Salesman Problem and its Variations*, Kluwer Academic Publishers, Boston, MA, 2018.

Shen, Kangshen; Crossley, John N.; Lun, Anthony Wah-Cheung: *Nine Chapters of the Mathematical Art, Companion and Commentary*, Oxford University Press, 2019.

Solin, Pavel: *Partial Differential Equations and the Finite Element Method*, Wiley-Interscience, 2015.

Stinson, Douglas R.: *Cryptography*, Discrete Mathematics and its Applications, Chapman & Hall/CRC, 2015.

Stoer, Josef; Bulirsch, Roland: *Introduction to Numerical Analysis,* Springer-Verlag, DE; New York, 2012.

Ward, J. P.: *Quaternions and Cayley Numbers*, Mathematics and its Applications, Kluwer Academic Publishers Group, Dordrecht, NL, 2017.

Index

❑❑❑